T3-BSE-261

Marketing and Pricing of Milk and Dairy Products in the United States

Marketing and Pricing of Milk and Dairy Products in the United States

Kenneth W. Bailey

IOWA STATE UNIVERSITY PRESS AMES

Kenneth W. Bailey received his PhD degree in agricultural economics from the University of Minnesota. Dr. Bailey was formerly a policy analyst for the U.S. Department of Agriculture and is presently Extension Associate Professor at the University of Missouri, Commercial Agriculture Program, Dairy Focus Team. His area of specialty is the marketing, policy, and financial analysis of the dairy industry.

♾ Printed on acid-free paper in the United States of America

First edition, 1997

Library of Congress Cataloging-in-Publication Data

Bailey, Kenneth W.

Marketing and pricing of milk and dairy products in the United States/Kenneth W. Bailey.

p. cm.

Includes bibliographical references and index.

ISBN 0-8138-2750-7

1. Milk trade—Government policy—United States. 2. Dairy products industry—Government policy—United States. 3. Milk—Prices—United States. 4. Dairy products—Prices—United States. 5. Agricultural price supports—United States. 6. Marketing orders—United States. I. Title.

HD9282.U4B28 1997 97-5474

381'.41371'0973—dc21

This book is dedicated
to the memory of
Zachary Martin,
who at the tender age of three
lost his battle with cancer.

We will miss him.

Contents

Part III / The Future of Milk Marketing

Preface

If you listen to the radio one morning and wonder why the local price of corn changed, you only need to know two things: basis, or the variation in local prices to central market prices like those of the Chicago Board of Trade, and market supply and demand. You might add in market psychology and government programs as other factors that may also affect prices. But by and large it is market forces that clearly and publicly determine market prices for corn.

But it's not that simple with milk. Dairy farmers don't really know what determines their milk prices. The fact is milk prices are largely determined by federal policies and programs, transportation costs, pricing policies of cooperatives, and overall supply and demand forces for milk and dairy products. And you may also add in overorder premiums, Class I utilization, effectiveness of tariff-rate quotas, the Dairy Export Incentive Program, state and local inspection programs, and state price control regulations. Milk pricing in the United States is a complicated business.

Dairy is the most regulated and complicated agricultural industry in the United States. Part of the reason is obvious: milk is highly perishable and has the potential to carry food-borne disease. Intensive and effective inspection programs in each of the states have made our milk supply safe and reliable. Another reason is that milk pricing policies have evolved over time and have been shaped by both legislation and court cases.

The current foundation for milk pricing was created out of marketing conditions during the first four decades of this century. Cooperatives sprang up to assist dairy farmers in bargaining with large proprietary processors in the 1920s. Then the Great Depression resulted in tremendous economic devastation to farmers. Federal milk marketing orders and the dairy price support program were developed to ensure an adequate income level for farmers and a wholesome supply of milk to consumers. In the intervening years court cases, the interstate highway system, and improvements in milk production, processing, and transportation technology all affected the milk pricing system.

The objective of this book is to clearly explain the major factors affecting milk pricing in the United States. A comprehensive and systematic approach is used to explain the basics of milk marketing. Complex subjects are broken down to their simplest terms and clearly illustrated using real-world examples. Once these subjects are mastered, their interactions are explored. In this way, the reader will gain a comprehensive understanding of how market forces, government intervention, and institutions interact to determine milk marketing in the United States.

Other reasons why milk marketing in the United States is complex are due to government regulations and policies and the diversity and market interactions of an assortment of dairy products. A thorough understanding of this subject is currently outside the grasp of but a few individuals with years of highly specialized training. This text will change that because it provides a thorough and systematic treatment of the major factors that determine milk marketing.

This book was written for persons with no previous training in the marketing of milk and dairy products. While no formal training in economics is required to read and understand major portions of this book, some exposure at the undergraduate level will help the reader understand the theory and practice of milk marketing. This book is the single source for all that a person will need to know to have a detailed understanding of how milk is marketed and priced in the United States.

The first section of this book provides a basic overview of the U.S. milk industry, including information on the historical need for government intervention in the market, where and how milk is produced, and the demand and varied uses for processed dairy products.

Chapter 2 reviews supply and demand for milk. The production of milk is briefly explained, from farm production methods to regional shifts in production over time. Seasonality of milk production and its effect on milk marketing are also discussed. The second part of the chapter deals with demand for milk and dairy products. There are two basic uses of milk from dairy farmers: for higher-valued bottled milk and for manufacturing purposes. The chapter also discusses trends in the use of milk for various dairy products and seasonality of demand. The chapter ends with a presentation of milk-equivalent conversions from pounds of dairy product to pounds of milk used for manufacturing.

Chapters 3, 4, and 5 present a brief review of major dairy products: fluid milk; soft manufactured dairy products such as yogurt, ice cream, and evaporated and condensed milk; and hard manufactured dairy products such as butter, cheese, and nonfat dry milk. This overview describes the properties of various dairy products, including their milk-equivalent conversions, industry standards of identification, consumption trends over time, and various end uses.

The second section of this book provides a detailed description of government policies and regulations and their effect on milk marketing and pricing in the United States. Each chapter provides a discussion of the origins and history of major policies such as dairy cooperative policies, federal milk marketing orders, the federal price support program, and local and state milk regulation and legislation. This is likely to be the most tedious section of the book, but it is essential to understanding milk marketing. Each chapter explains the origins and objectives for legislation and the mechanics of how each program works. In addition, detailed examples are given in order to provide a more thorough understanding.

Chapter 6 reviews the origins and functioning of federal milk marketing orders and clearly illustrates how they work. It provides a detailed discussion of classified pricing, pooling, calculation of blend prices, and how milk is marketed between orders. Chapter 7 analyzes the history of dairy cooperatives and their impact on milk marketing. Chapter 8 covers the federal dairy price support program and its relation to federal milk marketing orders. Chapter 9 discusses the history and development of local and state milk regulation, including sanitary milk regulations and state milk control and marketing orders.

The third section of this book focuses on the future of milk marketing in the United States, specifically international trade, and, in detail, discusses attempts at domestic and international policy reform.

Chapter 10 describes the international market for milk and dairy products. This chapter surveys the world market for dairy products, including major exporters and importers, U.S. import and export programs, and recent trade agreements such as the North American Free Trade Agreement and the Uruguay Round of GATT.

The last chapter summarizes milk marketing and pricing in the

United States and presents an objective assessment of government regulation of the U.S. dairy industry. Debate on the 1995 Farm Bill clearly focused attention on deregulation. One of the barriers to deregulating the U.S. dairy industry has been a lack of understanding of the complexities of current dairy policies. It is the intent of this book to shed light on an extremely complex policy regime. This chapter sorts out the pros and cons of deregulation and assesses dairy policy for the twenty-first century.

Acknowledgments

I would like to thank the following individuals who were so helpful to me in writing this book:

Rex Ricketts, coordinator of the Commercial Agriculture Program

Donald Nicholson, Agricultural Marketing Service

Bill Blakeslee and Roger Eldridge, Mid-America Dairymen, Inc.

Donald Kullmann, Prairie Farms Dairy, Inc.

Peter Vitaliano and Renee Selinsky, National Milk Producers Federation

Paul Christ, Land O'Lakes

Alden Manchester, USDA

Bob Cropp, University of Wisconsin

Milk Industry Foundation

Kelly Fisher and Martin Veeger, International Dairy Foods Association

Terry Long, Missouri State Milk Board

Evan Kinser, for help with the graphs

Diane Quintero, for typing numerous revisions of this book.

I would also like to thank my wife, Epifania, and two sons, Stephen and Travis, for missing many nights and weekends of family fun so that I could write this book.

Part I

The U.S. Milk Industry

Chapter 1

An Introduction to Milk Marketing

Milk is often called nature's most nearly perfect food. Campbell and Marshall (1975) attribute this first to Hippocrates, the father of modern medicine. No other food is as nutritious or has milk's reputation for purity and wholesomeness. Just the thought of milk conjures up images of red barns, white picket fences, and cows grazing contently in a pasture. Yet the reality of milk production, marketing, and pricing is in stark contrast to milk's image of simplicity: milk is often produced on large farms under factorylike conditions that in many cases no longer resemble the "family farm"; milk must be tested, processed, and packaged at a processing plant before it reaches the grocery store; no other commodity in the world is as highly regulated as milk; and no other commodity in the United States has been involved in as many legal challenges in regard to how it is marketed.

Americans are truly in love with milk and dairy products. Many Americans start their day with a bowl of cereal and a generous serving of milk. Children can buy milk in their school at reasonable prices thanks to the government-sponsored School Lunch Program. Milk is also highly nutritious. It contains 87 percent water and 13 percent solids. In its natural form, the solids portion of milk contains 3.7 percent milkfat and 9 percent solids-not-fat. The fat portion contains fat-soluble vitamins A, D, E, and K. The solids-not-fat portion of milk includes protein (primarily casein and lactalbumin), carbohydrate (primarily lactose), water-soluble vitamins, and minerals such as calcium, phosphorus, magnesium, and potassium. The calcium found in milk is readily absorbed by the body. The phosphorus helps the body to properly absorb and utilize this calcium. Milk is also a significant source of riboflavin (vitamin B^2), which helps promote healthy skin and eyes, as well as vitamins A and D (Milk Industry Foundation 1994).

What most Americans don't know about is the complexity of regulations that govern the production, marketing, and pricing of milk and dairy products. Most of these regulations are due to milk's perishability and capacity to carry food-borne diseases. Milk and dairy products must be inspected all the way from the cow's udder to the consumer's lips. Milk must be harvested from the cow and processed and distributed under strict sanitary conditions. Sanitary regulations have been developed and refined over the years at the municipal and state level in order to protect the public's health.

Evolution of Dairy Programs

While the need for sanitary regulations for milk and dairy products is obvious, it is less clear if government involvement is required to efficiently price and market them. Most of the state and federal regulations have their origins in the Great Depression. At the time, the economy collapsed, and consumers did not have the purchasing power to buy dairy products. As a result, demand fell, and milk prices plunged. Pricing plans that had been voluntarily worked out between milk cooperatives, milk processors, and dealers were no longer honored. Farmers panicked, and in many cases violence erupted as frustrated farmers confronted milk dealers and the public. Dairy farmers from California to New York staged milk boycotts in a failed attempt to control an increasing milk supply and to strengthen prices.

In order to combat low milk prices and the tremendous disruption in milk marketing, Congress and many statehouses enacted emergency legislation. Congress passed the Agricultural Adjustment Act of 1933 in order to help stabilize agriculture. This plan created a licensing system for milk marketing and for supporting milk prices. This was later modified into the federal milk marketing orders we know today. Commodity loans for butter became the precursor to the dairy price support program. Also during this time, many states enacted temporary legislation that in effect legislated price minimums at the farm, wholesale, and retail levels. Unprecedented legislation was required to help American farmers through an unprecedented collapse of the economy.

Despite the clear intent of Congress and state legislators to create temporary legislation to help farmers through the depression, many of the laws evolved into permanent legislation. After World War II the dairy industry began to experience tremendous change. Mom-and-pop grocery stores that sold milk came under increased competition from larger retail grocery store chains. These chain stores looked for large quantities of uniform quality milk and dairy products that were well suited for display and self-service (Williams et al. 1970). This in turn favored large-volume suppliers. Also, many of these grocery store chains used milk as a loss leader to entice shoppers into their stores. That effectively forced many of the mom-and-pop stores out of business. Milk processing also came under great pressure to change. Most communities across the country had one or two processing plants that bottled milk or processed dairy products. Many of these went out of business as larger bottling and manufacturing plants with greater economies of scale expanded. Small dairies that bottled milk from local milk supplies and delivered directly to the home were also forced out of business as consumers went to the larger grocery store chains to buy their milk. These chain stores purchased their milk from larger and more efficient fluid milk processing plants (known in the industry as "bottlers") that purchased milk from farmers over an increasingly larger geographical area. The advent of the interstate highway system and insulated milk trucks also allowed milk to travel greater distances from farm to plant to consumer.

As the general economy evolved, so the regulation of the dairy industry increased. Fluid milk for bottling needs was almost always supplied by local farmers. But as the feasibility of moving milk over greater distances became a reality, many state regulations were issued to keep outside milk from moving into local milk markets. Sanitary regulations were created to keep "foreign milk" out. State orders were modified to prevent pricing milk below cost. Federal orders were modified to prevent outside milk from benefiting from higher-priced local milk sales for bottling purposes. As the regulations increased, so did the court cases. Many cases were fought over the issues of milk pricing and milk movement.

Today there is a general trend to move the U.S. economy toward less regulation and free market principles in order to improve efficiencies in production and marketing and to lower costs to consumers. Yet despite this, the U.S. dairy industry continues to be governed by a number of complex regulations at all levels of government. There are two schools of thought here regarding the need for such regulations. Those who believe in government intervention argue that milk is a highly perishable product that cannot be efficiently marketed without direct assistance from the government. They point out that without government assistance, milk would be underpriced during periods of surplus production. They argue that federal orders were designed and modified over the years to create "orderly milk marketing conditions" and that without such direct government intervention milk prices would immediately collapse and the United States would return to the chaotic marketing conditions of the Great Depression.

On the other side, those who believe in free market principles argue that the economy has sufficiently changed to warrant a deregulation of the dairy market. The only farm commodities that are in surplus today are the ones that are supported by government programs. Prices would not collapse, and milk would be more efficiently marketed under an absence of government programs. Also, at the present time there is a relative balance between large national cooperatives and large proprietary dairy processors that did not exist during the turn of the century. Back then many small dairy producers were at the mercy of the large processor. Thus there is a more balanced bargaining situation between buyers and sellers today. In addition, the economy is sufficiently strong so that a milk price could be reached to effectively balance supply and demand under free market conditions. In short, those who favor a deregulation of the dairy industry argue that milk is no different than other farm commodities and would be more efficiency produced and marketed under free market conditions.

REFERENCES

Campbell, John R., and Robert T. Marshall. 1975. *The Science of Providing Milk for Man.* New York: McGraw-Hill Publishing Co.

Milk Industry Foundation. 1994. *Milk Facts: 1994 Edition.* Washington, D.C., September.

Williams, Sheldon, David A. Vose, Charles E. French, Hugh L. Cook, and Alden C. Manchester. 1970. *Organization and Competition in the Midwest Dairy Industries.* Ames: Iowa State University Press.

Chapter 2

The Supply and Demand for Milk and Dairy Products

The supply and demand for milk and dairy products have changed dramatically over the last few decades. Modern dairy production practices are employing embryo transfers to improve herd genetics, pedometers that track heat detection, and computers to balance rations, compute daily feed costs, and keep daily milk production and health records. Where milk is produced is also changing as milk production continues to shift regionally from the traditional areas of the Midwest and Northeast to the West and Southern Plains states. The demand for dairy products is also dynamic. Beginning in the 1970s, consumers rediscovered yogurt and rejected filled milk and artificial cheese, which substituted vegetable oil and casein for natural milk. Consumers clearly wanted the real thing. Interest also shifted to low-fat diets: consumers switched from whole milk to low-fat milk, and demand peaked for low-fat ice creams. Yet per capita consumption of cheese, a high-fat dairy product, continues to grow year after year, and consumption of high-fat premium ice cream represents a new growth area.

Supply of Milk

The nation's milk supply continues to increase each year as dairy producers learn to lower costs and expand output. In addition, growth in the U.S. population each year also results in an increase in consumption of milk and dairy products. In the past this has often created a surplus as supply increased faster than demand. This was particularly true in the 1970s and the first half of the 1980s when support prices were increased year after year by Congress in response to inflation and the political clout of dairy cooperatives. Excess milk had to be manufactured into storable dairy products and removed from the marketplace via government programs. Adding to the surplus problem was the adoption of new technologies that lowered production costs and led to greater efficiencies in milk production. Milk was gaining the reputation of being completely insensitive to market conditions as surpluses mounted.

The supply of milk today is more in alignment with the demand for dairy products. This resulted from a reduction in the support price of milk as well as other changes in government programs (Chapter 8). Lower supports led to greater volatility in the price of milk and less government removal of surplus dairy products from the marketplace. Dairy producers now adjust supply in response to general market conditions and are less affected by government support programs.

Milk Production

Milk production in the United States has grown from 127.5 billion pounds in 1961 to 155.6 billion pounds in 1995 (Figure 2.1). During the 1960s, annual production growth was negative in many years. This had policy makers concerned and contributed to efforts to raise the support price of milk (Chapter 8). During the 1970s and first half of the 1980s,

Figure 2.1. U.S. milk production

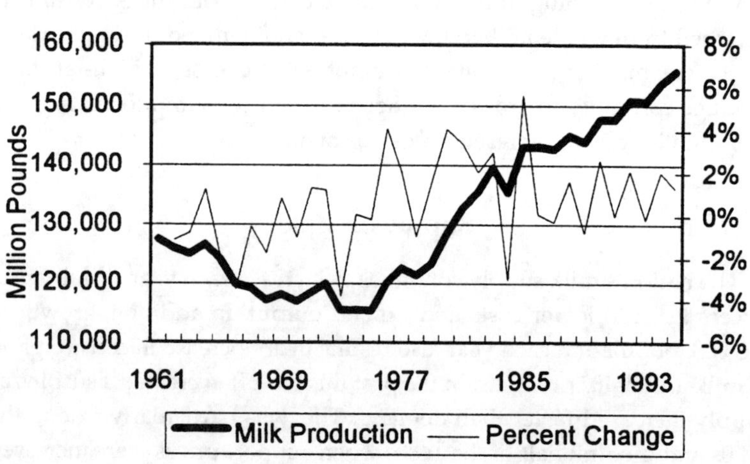

when government support to dairy farmers increased each year, milk production grew at an annual rate of 1 to 5 percent (excluding off years and the Milk Diversion Program in 1983). Then, during the latter half of the 1980s and first half of the 1990s, production growth stabilized (due to reduced support prices) and grew in a narrower range of zero to 2 percent.

Production per cow has grown steadily since the end of World War II and particularly since the 1960s, when new production technologies began to be used by dairy farmers. The Dairy Herd Improvement Association began a widespread program to develop production databases to improve dairy herd management practices. Herd genetics improved via widespread use of artificial insemination. Feeding practices changed as producers learned about the correlation between forage quality and milk production and began to feed concentrates outside the parlor (where cows are milked). Better parlor and housing designs were used to improve cow movement and comfort and to better use scarce labor resources. The result was a gradual increase in milk sales per cow, from 7,326 pounds per cow in 1961 to 16,451 pounds in 1995 (Figure 2.2). This linear increase year after year represented annual growth rates ranging from −1 to 5 percent. It narrowed in the last 10 years to about 2.3 percent each year, or almost 300 pounds of milk per cow per year.

Figure 2.2. Milk sales per cow

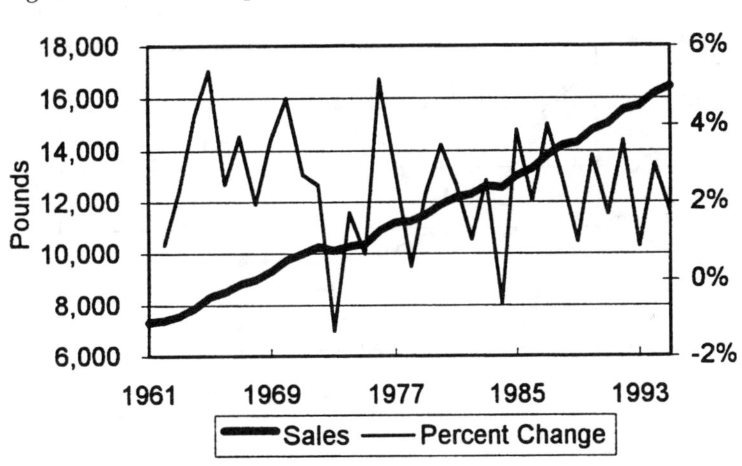

While part of the annual growth in milk sales per cow could be attributed to better management and technology, some could be attributed to a reduction in the number of dairy farmers. High-cost dairy producers with heavy debt loads were forced out of business over time. As those producers left the industry, national average milk sales per cow increased. This is because those producers also tended to be less efficient and realized below-average production per cow. Those producers sold some of their production resources (namely their better-quality cows) to the remaining more efficient producers. This in essence represents Cochrane's treadmill theory, which explains the concentration of production resources in agriculture since the end of World War II.

Fewer cows were needed each year to meet market demand due to improved sales per cow. As a result, cow numbers declined from 17.4 million head in 1961 to 9.5 million head in 1995 (Figure 2.3). There was a tremendous decline in cow numbers in the early 1960s, particularly in the Midwest. Cow numbers fell sometimes 4 to 7 percent from the previous year during this period. Many dairy farmers left the industry due to low profit margins and better opportunities elsewhere (e.g., grain farming). It was during the 1960s that increased sanitary regulations forced many dairy farmers to switch from spring-cooled milk storage systems that used milk cans to refrigerated bulk tanks that stored a one-

Figure 2.3. Cow numbers on farms

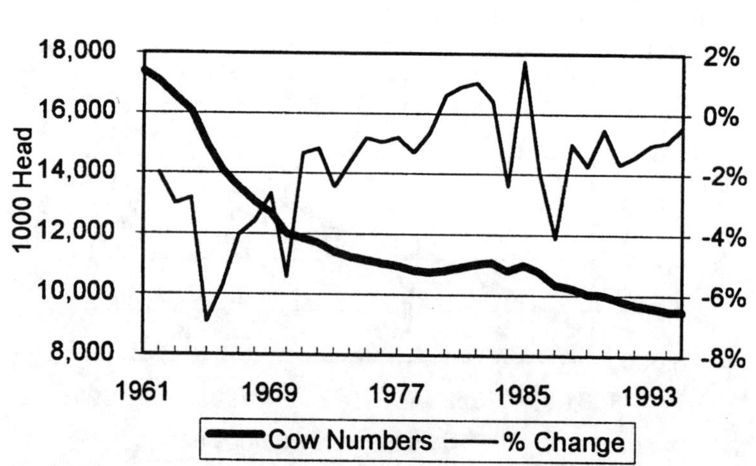

to two-day supply of milk. Many smaller producers simply could not afford to make this important transition. Also, the gap in earnings between producers who met the higher standards of Grade A milk production and those who did not (usually called "Grade B producers") was such that many Grade B producers either upgraded their facilities to meet Grade A standards or left the industry. Milk hauling routes for Grade B producers were in some cases eliminated as smaller creameries gave way to a general consolidation of dairy processing plants.

This strong decline in cow numbers stabilized and then reversed itself in the latter half of the 1970s and early 1980s. The rise in the support price for milk coupled with a reduction in grain prices resulted in many farmers entering dairying or expanding their operations. Cow numbers actually increased from 10.7 million head in 1979 to 11.1 million head in 1983. Milk support prices peaked at $13.49 per hundredweight in 1981. Thereafter, government programs such as the milk diversion and whole herd buyout programs (Chapter 8) and a general reduction in the milk price support level contributed to a reduction in cow numbers.

Milk is produced in a fairly uniform seasonal pattern in the United States (Figure 2.4). While much less pronounced today than, say, 15 years ago, milk production usually peaks in May due to spring weather, green pastures, and improved forage quality. Also, many cows calve in early

Figure 2.4. Daily average milk production in 22 select states

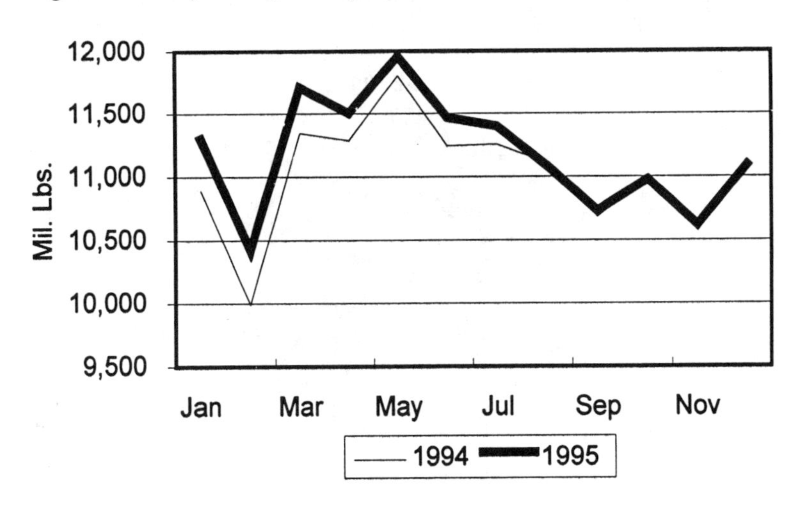

spring and begin a new lactation cycle. Daily milk production is greatest during the early portion of a cow's lactation cycle. National milk production begins to fall off significantly during the summer when hot and humid weather adversely affects forage quality and milk production. Cows are particularly vulnerable to heat and humidity and typically don't eat as much as in cooler climates. Less feed translates into lower milk production. Production begins to pick up again in the fall when other cows calve and temperatures begin to cool.

Size of U.S. Dairy Farms

According to the National Agricultural Statistics Service, about 42 percent of the nation's supply of milk was produced on farms with fewer than 99 head of cows in 1994 (Figure 2.5). Fewer farmers and fewer cows have over time led to an increase in the size of remaining dairy farms. While one family could profitably milk 50 head of dairy cows in the 1970s and 1980s with a moderate to heavy debt load, that is less true in the 1990s.

Much of the recent growth in the size of dairy farms is due to what economists call "economies of scale." In other words, it is cheaper to produce a unit of milk as farm size increases. There are three factors that contribute to economies of scale in milk production: investments, labor,

Figure 2.5. Percentage of milk production by size of farm

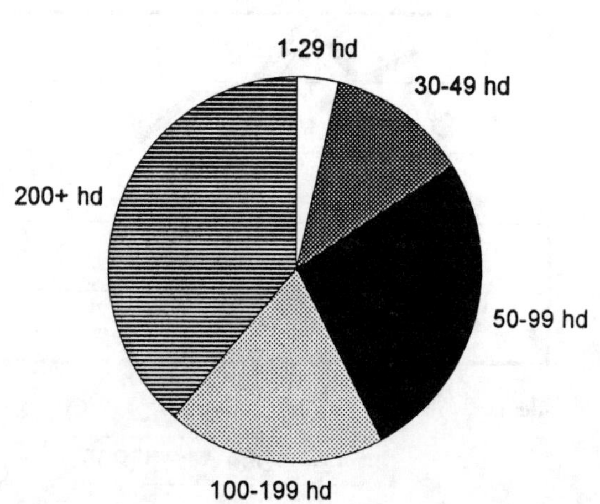

and variable inputs. Dairy farms are very capital-intensive due to the need to produce milk in confinement and to meet strict sanitary conditions. Large investments must be made in the milk house and milking equipment, feeding systems, cow housing, waste management, and farm equipment. In general, the larger the number of cows, the lower the total investment per cow on a modern dairy farm. There are more cows on which to spread the initial investment in real estate, housing, machinery, and equipment. Also, larger dairy farms use fewer labor inputs per cow or per unit of milk production. Fewer hands are required on a large modern dairy farm with an efficient parlor and modern cow housing with an automatic flush system to handle dairy waste. Larger farms can also purchase parlor supplies and other variable inputs at cheaper prices than smaller farms. They get discounts for purchasing large amounts of feed, and volume premiums and hauling discounts for marketing large volumes of milk.

All of these factors result in larger farms having the ability to produce milk at a cheaper price if their management can take advantage of economies of scale. Dairy, unlike other agricultural enterprises, is very sensitive to management. Two farms side by side with similar investments may have surprisingly different production levels due to differences in management. That is because dairy cows are very sensitive to their environment. Differences in how many times per day a cow is milked, availability and quality of feed, reproductive management, and general cow comfort will result in very different levels of milk production.

Production Systems in the United States

The location of milk production has also changed dramatically in recent years. This change is primarily because of economies of scale and differences in management styles and production methods. Production is growing in the Southwest and western regions of the United States on a percentage basis, but it is shrinking in the Upper Midwest and Northeast. This reversal likely results from the growth of large western dairy operations, some as big as 1,000 to 15,000 cows. Additionally the western style of management focuses on the dairy operation, hires and effectively manages labor, and employs financial and production records to manage the operation.

Before examining the regional shifts in milk production, it is useful to understand the alternative ways milk is produced on dairy farms in the United States. At present, there are four major methods of milk production in practice. The first approach centers around tie stall or stanchion barns, wherein cows are tethered, milked, and bedded in one stall. A central milk pipeline runs the length of the barn. During milking time, the person milking the cows moves the milking unit from stall to stall. Feed is brought directly to the individual cow. Hay is often stored upstairs in a hayloft. In addition, an upright silo where hay and corn is stored and available for feeding is usually built near the barn for ease of access during inclement weather. This type of system is still employed in many parts of the Upper Midwest and in the Northeast because of the need to limit trips to and from the barn during the cold winter months. There are three major problems with this system:

1. It is very labor-intensive since cow stalls must be cleaned and bedded, individual cows must be hand-fed, and milking is laborious.
2. This approach is very costly and requires a large investment on a per cow basis. It typically ranges from $7,000 to $9,000 per cow (including costs of land and the cow).
3. Expansions are difficult and limited since adding more cows requires a new barn, milking equipment, and stalls.

This system is often used on small dairy farms with less than 100 cows.

The second production system, called "the traditional approach," separates the milking facilities and housing. These operations typically have less than 100 cows and employ family labor as opposed to hiring most of the labor needs. Cows are milked in a parlor and are housed in some type of confinement facility. There are many types of parlors that are in use today. These range from flat barn parlors, which require the milker to bend over to milk the cow, to four-on-a-side operations, where cows stand end to end to be milked, to double-sided herringbone parlors, where the milker stands in a pit and cows are angled to the pit. A recent trend is toward rapid exit parallel parlors.

Housing is often used with the traditional approach, particularly in colder climates. This housing may range from concrete corrals with feeding troughs and Utah-type stalls, to free stall housing. Utah housing is essentially stalls with a roof overhead. This type of system was very popular in the 1970s and 1980s as it required very little investment, but it had limited protection for cows and feed from the elements. Free stall housing includes a large barn with dry, comfortable stalls that cows can choose to lie and rest in. Feeding areas are typically located either down the middle of the housing, or on one or both sides. In the traditional system, cows are usually fed concentrates (a pelleted ration consisting of, for example, corn, soybean meal, and vitamins and minerals) in the parlor, and forages (hay, haylage, and silage) in either an outdoor lot or in the free stall barn. The trend today, however, is to feed outside the parlor and to mix concentrates and forage together and feed at the same time. This approach is called a "total mixed ration."

Investments in a traditional dairy farm may average $4,000 to $7,000 per cow (including the land and the cow).

Both the tie stall and traditional approaches to dairying typically include cropping enterprises as part of their business practices. Depending on location, accessibility to market sources of consistent quality forages, and production practices, the traditional dairy farmer often relies on cropping enterprises to make effective use of manure and to supply quality forage for the dairy. Quality forage is essential to milk production, and smaller dairy farms often do not have the ability to purchase quantities of consistent quality alfalfa hay due to their size of operation. Also, many dairy producers prefer to diversify their operations because their families traditionally have. They employ more family labor by raising row crops and by managing beef and hog operations.

Modern milk production systems take advantage of innovations in parlor management, computerized records, efficient feeding systems, and hired labor to handle greater cow numbers and to expand production. There are two approaches within the modern milk production system: free stall housing and drylot operations. Both approaches use modern parlors that require just one to three milkers (depending on size), improve cow flow in and out of the milking stalls, and employ computerized equipment that can measure individual cow production each milking.

Modern free stall operations house cows in free stall barns that maximize air flow and ventilation, allow for total mixed ration feeding systems, and use flush systems to remove dairy waste. These types of dairy operations can be found in all regions of the United States. Ventilation is critical to cow comfort and milk production. Hot stuffy barns result in a direct reduction in milk production. A total mixed ration combines both concentrates and forages into a single mixed ration. As a result, each bite the cow takes has a well-proportioned mixture of both concentrates and forages. This diet improves both cow nutrition and rumen health, which is critical to milk production. Flush systems are used to remove waste from the parlor, walkways, and free stall housing with little or no labor. Thus flush systems reduce labor costs, improve cow comfort, and result in cleaner udders and better-quality milk. Total initial investment in a dairy with a modern free stall operation typically runs from $3,300 to $3,600 per cow (including the costs of cow and land).

Modern drylot operations are used in the southwestern portion of the United States in arid and semiarid climates. Cows are confined outdoors in lots, or paddocks. Mud is not a problem in these paddocks if the operation is located in an arid part of the United States. Dairy waste is regularly scraped into large piles and hauled away once or twice a year. Cows are moved into the milking parlor two or three times per day to be milked. Drylot operations are usually large in order to take advantage of economies of scale. Total initial investment in a modern drylot operation (including the costs of cow and land) is about $3,000 per cow. Investments in free stall housing, which typically run about $850 per stall, are not made. Thus significant savings in capital and debt costs are realized.

A new approach that is being used on many smaller dairy operations in the United States is rotational intensive grazing, or "grazing" for short. This approach originated in New Zealand, where cows are grazed on high-quality pastures and are usually not fed expensive concentrates. In grazing systems, paddocks (cells of pasture) are developed. Cows are then rotated from paddock to paddock. These paddocks are well managed in order to provide a quality forage for the lactating dairy cow. The advantage of this approach is that forage quality is increased, concentrate costs are usually lowered, and housing and machinery costs (for harvesting hay) are usually lowered. The disadvantage in the United States is that often a dual system is needed: housing and hay harvesting, both at a cost. Also, the approach is often not feasible in environments where mud is a problem.

Regional Milk Production Shifts

Milk production has been shifting to less traditional dairy regions of the United States due to (1) increased competition for a limited domestic market and (2) large-scale dairy operations learning to improve profits via economies of scale. Less milk is produced today in the traditional dairy regions of the Lake States, Northeast, and Corn Belt, and more is now produced in the Southern Plains, Mountain, and Pacific regions (this is based on the percentage of national milk production; see Table 2.1). Over the period 1977–94, U.S. milk production expanded 23 percent. During this period some states expanded production while others reduced it. Western and Southern Plains states from Washington to Texas showed above-average gains in production. New Mexico was the fastest-growing state, up 515 percent. Many Midwest and Northeast states, however, either expanded at below-average rates or showed declines in production.

California, now the largest milk producing state in the United States, has seen a significant expansion in cow numbers over the years. Dairy farms there tend to be large, usually 700 to 1,500 cows. In recent years dairy producers in California have been relocating within the state and

Table 2.1. Percentage of regional share of U.S. milk production

Region	1965	1975	1980	1985	1990	1993	1994
Northeast	20.7	20.4	20.4	20.0	18.3	18.6	18.1
Lake States	28.3	28.0	28.7	28.7	26.7	25.3	24.3
Corn Belt	17.1	13.6	12.4	11.8	11.5	10.9	10.4
Northern Plains	5.3	4.6	4.1	3.9	3.6	3.2	3.0
Appalachian	6.9	6.9	6.6	6.1	5.6	5.3	5.0
Southeast	3.0	3.8	3.5	3.1	3.3	3.3	3.3
Delta	2.3	2.3	2.0	1.8	1.7	1.6	1.6
Southern Plains	3.5	3.7	3.7	3.6	4.6	4.7	4.9
Mountain	3.7	4.4	4.8	5.5	6.4	7.5	8.5
Pacific	9.2	12.3	13.8	15.5	18.3	19.7	20.9

Source: USDA, ERS 1995.

have also been migrating to other states due to overpopulation in California and land appreciation. These farmers have often moved with large amounts of capital after selling off farms located adjacent to urban centers. They have moved to states like Texas and New Mexico and have started large modern drylot operations. Other states such as Washington, Oregon, and Idaho have also seen expansions in cow numbers and milk production in recent years. These expansions have often involved 1,000 to 5,000 cows. New Mexico, for example, has seen a large increase in milk production, up 27 percent in 1994 from the year before. Almost all of this increase was from investments in new dairy operations. One of the problems with such rapid expansion has been finding a local market for the milk. New Mexico has a limited population base and thus a limited possibility for consuming milk in fluid form. Until recently it also has had limited capacity for manufacturing surplus milk. Production has also increased in Texas, thus limiting a marketing opportunity in New Mexico. As a result of all these factors, significant quantities of milk have had to be hauled outside the state, thus increasing transportation costs for which dairy farmers have had to pay. Milk processing capacity, however, has been expanding in the state.

Demand for Milk

Milk is consumed in the United States in a variety of forms, either as fluid milk in packaged containers or as a manufactured dairy product such as butter, cheese, or nonfat dry milk. There are numerous products that are manufactured from dairy products. Some products, such as cheese, ice cream, and yogurt, are consumed as a final processed product. Others, such as nonfat dry milk, are used as an ingredient in other foods such as candy bars and bakery goods.

In general, milk is utilized for three major purposes: (1) fluid products, (2) manufactured products, and (3) on-farm use. In 1993, 36.6 percent of the milk supply was used for fluid products, 60 percent for manufactured dairy products, and just 1.3 percent for on-farm use (the balance was unaccounted for; see Table 2.2). The level of fluid milk consumption is critical in the calculation of milk prices since farmers receive a blend of both fluid milk and manufactured milk prices. Federal and state orders place a higher value on milk consumed in fluid form. On average, the difference in value between milk used in the bottle and milk

Table 2.2. Utilization of milk in the United States, 1993

	Million lbs.	Percentage
Milk used in manufactured products		
Creamery butter, total ME	29,493	
ME of butter from whey cream	4,500	
Net milk equivalent	24,993	16.5
Cheese		
American	29,415	19.5
Other	20,456	13.5
Cottage cheese, creamed	559	0.4
Total milk used for cheese	50,430	
Canned milk		
Evaporated, sweetened condensed	1,178	0.8
Bulk condensed whole milk:		
Unsweetened	374	0.2
Sweetened	324	0.2
Dry whole milk	1,130	0.7
Ice cream and other frozen dairy products, total ME	14,058	
ME of butter and condensed milk used in ice cream	1,995	
Net milk equivalent	12,063	8.0
Other manufactured dairy products	199	0.1
Total manufactured dairy products	90,691	60.0
Milk available for use in fluid products		
Sold by dealers	54,361	36.0
Sold by producers directly to consumers	968	0.6
Total available for fluid products	55,329	36.6
Milk used where produced		
Fed to calves	1,450	1.0
Consumed as fluid milk, cream, and farm-churned butter	446	0.3
Total milk used by producers	1,896	1.3
Residual	3,160	2.1
Total Utilization	151,076	100.0

Source: USDA, NASS 1994.

Note: ME represents milk-equivalent conversions on a butterfat basis.

used for manufacturing purposes is over $2 per hundredweight. As a result, milk is first used in bottled form and the balance used for manufacturing purposes. Thus, if fluid milk consumption is depressed in a particular year, the balance is manufactured into lower-valued dairy products. The impact on farm-gate milk prices is to lower the average blend price.

Adding to the complexity of milk marketing is the fact that fluid milk consumption is nearly inversely related to milk production over the course of a year (Figure 2.6). In other words, fluid milk sales normally bottom out in late spring and early summer when children are out of school and milk production normally expands. Demand then peaks for the year in the fall months when children return to school and when milk production normally bottoms out. This exacerbates the problem of milk marketing since Grade A quality milk must be made available to a market all year round. Thus, the optimum level of supply for a local market will be such that supplies are adequate to meet fluid needs in the fall months. If reached, this level of production will result in excess milk production in the spring months, which will need to be processed into manufactured dairy products. But a strong demand for manufactured dairy products will ensure that surplus Grade A milk will be competitively priced. Prices are theoretically set under federal orders in order to en-

Figure 2.6. Milk supply versus fluid milk consumption

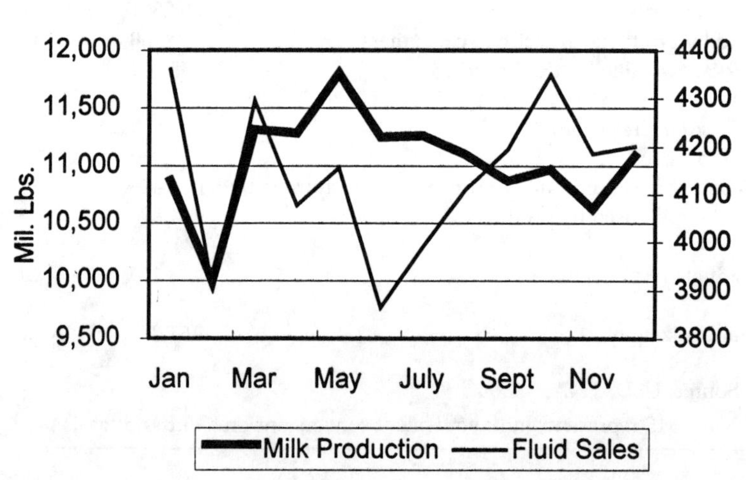

courage overproduction so that supplies of fluid-eligible milk will be available all year round.

Most milk used for manufacturing purposes is processed into storable dairy products: butter, dry milk products, and cheese. In 1993, of all milk used for manufacturing purposes, 56 percent was used to produce cheese (American, other hard cheeses, and cottage cheese), 28 percent was used for butter and powdered milk products, 13 percent was used in ice cream, and 3 percent for canned milk products. Under federal and most state orders, cheese represents the next best alternative use for milk in terms of market value. If cheese consumption is depressed in any one year, the resulting surplus will likely be manufactured into the lowest-value use—for butter and powdered milk. Thus milk used for butter and powder production represents a residual class for milk not used for fluid or cheese production. Many dairy cooperatives own butter/powder production facilities and use them more heavily during surplus periods of the year when milk production in excess of immediate needs has to be processed, or dumped.

Factors Affecting the Demand for Milk

Describing the factors that affect the demand for milk is complicated in part because of the varied number of products that can be processed from milk and consumed. Factors that influence why one consumer would purchase skim milk are considerably different from factors that affect another's decision to buy cheese. In addition, dairy products are purchased in three different ways: (1) retail sales for at-home use, (2) institutional sales (to hospitals, prisons, the armed forces, etc.), and (3) restaurant sales.

Economic theory states that demand for a particular product is a function of the price of the good, the price of a close substitute, income, and other socioeconomic and demographic factors (Haidacher 1992). Thus dairy products should be inversely related to their price. In the real world, however, small changes in price usually have little bearing on consumption of many dairy products (e.g., fluid milk and cheese). In addition, it is difficult to isolate close substitutes for many dairy products. While margarine is clearly a close substitute for butter, there is no obvious substitute for cheese and fluid milk. The Economic Research Service lists three general factors that influence the demand for milk and dairy products. They are (1) population, (2) the civilian unemployment rate, and (3) per capita real personal disposable income. Generally, as the population increases, so does consumption of dairy products. In order to isolate the impact of population growth on commercial sales of dairy products, economists often divide total consumption by population in order to arrive at per capita sales. Thus changes in consumption are often measured in terms of per capita sales. The unemployment rate is also a key factor affecting the demand for milk and dairy products. As the unemployment rate declines, consumption of dairy products should increase. When people are employed, they have the buying power to purchase dairy products for personal consumption, for home entertaining, or away from home in restaurants. The same is true with an increase in personal disposable income. Disposable income is discretionary income that can be spent on milk and dairy products or on other competing products.

The rationale for using socioeconomic and demographic factors—such as age, sex, race, family size and composition, and education—is that they represent nonprice factors that may significantly affect consumption. But economists often find these difficult to measure. Other factors that may affect the demand for dairy products are government policies, such as domestic donation programs and food stamp programs, and advertising and promotion programs.

The National Milk Producers Federation (1995) recently completed a study of the factors that affect the demand for various dairy products. The results of that study are summarized in Table 2.3. The columns represent the dairy products a model was estimated for. The rows indicate the various factors that were used to estimate the demand function. The cells in the table present two pieces of information: (1) the relation between the demand for a particular dairy product and the demand factor and (2) the statistical significance of that factor.

For example, when estimating per capita demand for fluid milk products (columns 1–4 in Table 2.3), one would expect an inverse relation between demand and the price of milk, and a positive relationship between demand and income. The results of this study indicate that, in general, a rise in the price of milk has a modest impact in lowering the demand for milk, and that a rise in consumers' income (in this case per capita personal income) has a modest impact in increasing the demand for milk. The level of education for the general population and the percentage of

Table 2.3. Selected economic and demographic factors that impact the demand for dairy foods

	Fluid milk products				Cheese				Frozen
	Total	whole	Low-fat	Skim	Total	American	Other	Butter	dairy
Own price	negative modest	negative medium	N/S	positive medium	N/S	N/S	N/S	negative medium	negative modest
Substitute price	—	—	—	—	—	—	—	N/S	N/S
Income	positive modest	positive medium	N/S	N/S	positive large	positive large	positive medium	N/S	positive modest
Education	positive medium	N/S	positive large	N/S	positive large	positive large	positive large	negative large	N/S
Percentage U.S. population 0–15 years old	positive medium	positive medium	N/S	positive large	positive medium	positive large	negative medium	negative large	—
Percentage U.S. population 30–49 years old	—	—	—	—	negative large	negative large	negative medium	—	—
Percentage U.S. population 65+ years old	—	—	—	—	positive large	positive large	positive large	—	—
Percentage U.S. population nonwhite by race and ethnicity	negative large	negative large	N/S	negative large	positive large	N/S	positive large	positive medium	—
Percentage women in the workforce	—	—	—	—	positive large	positive large	negative medium	—	—

Source: "Trends in U.S. Dairy Demand" 1995.
Note: N/S = not statistically significant; — = not estimated.

the population between the age of birth to 15 years old have a medium effect in increasing the demand for milk. The ethnicity of the U.S. population (the percentage of the U.S. population that is nonwhite) has a significant negative impact on the demand for fluid milk.

The demand for cheese does not appear to be affected by the price of cheese. In other words, an increase or decrease in the price of cheese does not have a statistically significant impact on cheese consumption. The consumers' level of income, however, has a large to medium impact on the demand for cheese. Also, the older the population, the greater the ethnicity of the population, and the larger the number of women in the workforce all result in increased demand for cheese.

The demand for butter is inversely related to the price of butter. In other words, as the price of butter increases, butter consumption is reduced. The price of margarine, a substitute for butter, does not appear to impact the demand for butter during the period of study. The higher the median level of education for the general population and the greater the number of 0- to 15-year-olds both appear to reduce butter consumption. This suggests that educated people and youngsters don't consume butter. But the degree of ethnicity of the U.S. population does appear to have a positive impact on butter consumption.

Finally, frozen dairy products are negatively related, to a modest degree, to the price of frozen dairy products. Income, however, has only a modest positive impact on consumption.

This study clearly suggests that socioeconomic and demographic factors will have a greater impact on the demand for milk and dairy products in the United States as the median level of education, age, ethnicity, and women in the workforce increases.

Milk-Equivalent Conversions

Dairy market analysts track the supply and demand for milk to assess market conditions, policy, and the impact of trade on the domestic market, among other things. The USDA's supply and demand balance sheet for milk and all dairy products is presented in Table 2.4. The nation's milk supply is derived by adding imports and beginning stocks to domestic production. But both imports and beginning stocks represent dairy products, not volume of raw milk. Both imports and beginning stocks of milk in this table are represented via milk-equivalent conversions (used to estimate the pounds of milk used to process or manufacture dairy products). Also, the demand for raw milk is estimated by adding domestic commercial disappearance of milk to exports and shipments. (Commercial disappearance is a residual category computed in the supply and demand balance sheet after surveying imports, production, exports, and stocks. Commercial disappearance is a proxy for consumption.) Pounds of dairy products removed from the marketplace by the government through the dairy price support program or government-assisted export programs are subtracted from this total. Once again, domestic commercial disappearance, export and shipment, and government removal figures represent processed dairy products, not pounds of raw milk. The numbers in Table 2.4 were derived by converting product pounds into the equivalent amount of raw milk used to manufacture these products.

Table 2.4. Milk: supply and utilization of all dairy products (in million pounds), 1990–94

	1990	1991	1992	1993	1994
Supply					
Production	147,721	147,697	150,885	150,582	153,622
Imports	2,690	2,625	2,521	2,806	2,880
Beginning stocks[a]	9,036	13,359	15,840	14,214	9,570
Total supply	159,447	163,681	169,246	167,602	166,072
Utilization					
Fed to calves	1,484	1,480	1,436	1,408	1,353
Human use	142,067	142,897	144,519	148,178	152,791
Exports and shipments[a]	2,537	3,464	8,147	8,446	6,168
Total use	146,088	147,841	155,032	158,032	160,312
Ending stocks[a]	13,359	15,840	14,214	9,570	5,760

Source: USDA, ERS 1995.

Note: The supply and demand figures for dairy products were determined on a milk-equivalent, milkfat basis. In other words, the numbers were derived by converting product pounds into the equivalent amount of raw milk used to manufacture these products.

[a]Government and commercial.

Converting from pounds of processed dairy products back to raw milk may at first glance appear simple. One could use the pounds of milk required to produce a pound of cheese, right? But the problem with a straight product conversion using yield factors (e.g., it takes 100 pounds of milk to make 4.48 pounds of butter) is that the production process for many dairy products also produces other by-products. Some of these by-products are further used in other manufactured dairy products. Cheese, for example, produces whey solids, which are refined into dry whey, lactose, and whey butter. Butter production jointly produces nonfat dry milk and buttermilk as by-products. Therefore, directly converting pounds of dairy products back to raw milk using yield factors may grossly misrepresent the total amount of raw milk consumed by the marketplace.

A preferred way to convert from product pounds to raw milk equivalents is to use milk-equivalent (ME) conversion factors that take into consideration the components available in milk. These factors are available on a product-by-product basis. For example, there is one ME conversion factor for American cheese and another for lactose. ME conversion factors convert product pounds of various dairy products into milk-equivalent units based on the composition of milk in relation to the products converted. Two of the most common ME conversions are on a milkfat and a solids-not-fat (SNF) basis. Raw milk on average consists of 3.67 percent milkfat and 8.60 percent SNF. The SNF component consists of protein, lactose, ash, and other minerals. Therefore 100 pounds of milk on average will contain 3.67 pounds of milkfat, 3.20 pounds of protein, 4.75 pounds of lactose, 0.65 pounds of ash and minerals, and 87.73 pounds of water (Jacobson 1992). An ME conversion factor on a milkfat basis, for example, will convert product pounds into pounds of milk based on the relative proportion of milkfat in the individual dairy products and in raw milk. It compares the percentage of milkfat in cheese, for example, with an average 3.67 percent milkfat found in raw milk. This is the most commonly used ME and was used in the construction of Table 2.4. An ME conversion factor on an SNF basis will convert product pounds into pounds of milk based on the SNF composition of a particular dairy product to an average 8.60 percent SNF composition of raw milk. Other ME conversion factors can be developed based on calcium, lactose, and ash (Campbell and Marshall 1975). In essence, ME conversion factors are used to calculate the amount of milk components found in dairy products and to convert these components into the pounds of raw milk they represent.

To see how ME conversion factors work, let's look at an example of how a milkfat ME conversion factor for butter would be developed. This conversion factor takes into account the fat content of both butter and raw milk. Butter contains on average 80 percent milkfat and 1 percent SNF. Therefore, the ME conversion factor for butter on a milkfat basis would be calculated as follows:

Butter, ME conversion factor, milkfat basis	=	Percentage milkfat in butter/percentage milkfat in raw milk
	=	80.0/3.67
	=	21.80

The USDA reported that in 1993 commercial disappearance for butter was 1,040.4 million pounds. How much raw milk does that represent? Multiplying 1,040.4 by the ME factor for butter of 21.80 yields a milk equivalent of 22,680.7 million pounds of raw milk used to produce the butter consumed in 1993.

For another example, let's look at the ME of dry whole milk (DWM). DWM consists of 71 percent SNF and 26.5 percent milkfat. Therefore, the ME conversion factor for DWM on a SNF basis would be calculated as follows:

Dry whole milk, ME conversion factor, SNF basis	=	Percentage SNF in dry whole milk/percentage SNF in raw milk
	=	71/8.60
	=	8.26

The calculation of the ME factors are not always as simple as these examples suggest because of the problem of accounting for residual uses of dairy products. For example, cheddar cheese contains on average 33.14 percent milkfat (USDA, ARS 1976). Dividing 33.14 by 3.67 (the percentage of milkfat in raw milk) yields 9.02. But the USDA uses an official milkfat ME of 9.23 for cheddar cheese in order to calculate the ME of dairy products removed from the marketplace by government programs

(USDA, ASCS 1990). The reason is that this conversion factor has been adjusted upward to account for milkfat contained in the by-products whey fat and dry whey. Using a milkfat ME conversion factor of just 9.02 would understate the amount of raw milk used to process a pound of cheese.

Another problem is nonfat dry milk. According to the USDA, nonfat dry milk contains on average 96.07 percent nonfat solids (USDA, ARS 1976). Dividing 96.07 by 8.60 (average SNF composition of raw milk) yields 11.17. But the USDA uses an SNF ME of 11.58 for nonfat dry milk (USDA, ASCS 1990). Here again, the USDA has adjusted the SNF ME conversion for nonfat dry milk in order to account for the SNF contained in butter and buttermilk, two by-products of nonfat dry milk production.

Another ME conversion factor used is based on the total solids in dairy products (the combination of both milkfat and SNF). It is simply a weighted average of the milkfat and SNF ME conversion factors. In the Food, Agriculture, Conservation, and Trade Act (FACTA) of 1990, Congress requested the USDA to estimate the milk equivalent of dairy products removed from the marketplace by government programs on a total solids basis. The ME calculation of government removals was an important calculation for dairy producers since triggers built into FACTA to adjust the milk support price and to determine producer assessments were based on this calculation (Jacobson 1992). Prior to FACTA, government removals were always estimated on a milkfat basis. The reason for the change was because butter and cheese removals (both high in milkfat) prior to FACTA greatly exceeded nonfat dry milk removals. That in effect overstated the ME of government removals on a milkfat basis. FACTA required that the weight for milkfat in the calculation of the total solids ME conversion factors could not exceed 40 percent and the weight for SNF could not exceed 70 percent. The USDA adopted a 40/60 weighting between milkfat and SNF ME conversion factors based on FACTA in order to have a more balanced method of accounting for surplus removals of dairy products under government programs. So, for example, an ME for butter would be 21.80 on a milkfat basis and 0.12 on a SNF basis, but would be 8.79 on a total solids basis. Jacobson notes that on average, raw milk contains 12.27 pounds of total solids, of which only 30 percent (3.67 pounds of milkfat/12.27 pounds of solids) is accounted for by milkfat. Congress in effect chose to adopt a total solids weighting that actually overstates the average fat content of raw milk.

At the present time there are only two official sources of ME conversion factors available from the USDA. One USDA source estimated the ME conversion factors for dairy products removed from the marketplace by the dairy price support program (butter, nonfat dry milk, cheese, and evaporated milk) (USDA, ASCS 1990). ME conversion factors were estimated on a milkfat, SNF, and total solids basis. A second USDA source estimated ME conversion factors for imported dairy products in order to establish market access requirements under the Uruguay Round of GATT (General Agreement on Tariffs and Trade).

The only complete set of ME conversion factors available for all dairy products has been computed by the National Milk Producers Federation. Its set of ME factors incorporates the published USDA numbers and estimates the rest from USDA Handbook 8 (USDA, ARS 1976). This set of ME conversion factors is presented in Tables 2.5 and 2.6. Table 2.5 contains ME conversion factors for imports, and Table 2.6 for domestic consumption, government removals, and exports. The reason for two different sets of conversion factors is that the standards of identity (the milkfat and SNF contents of processed dairy products) are different for imported dairy products than for domestically produced products.

A few caveats are in order before using the ME conversion factors supplied by the National Milk Producers Federation. First, there are clear differences in standards between U.S. and imported dairy products. For example, the fat content in imported ice creams is often higher than in domestically produced ice creams. The estimates provided in Tables 2.5 and 2.6 attempt to account for these differences. Second, the MEs provided were estimated for a few established categories of dairy products. MEs for individual products may differ from those for the product category due to differences in standards of identity and manufacturing procedures. Third, the MEs presented may not eliminate all double counting for total supply and utilization of milk. Examples of problem areas are butter and condensed milk used in ice cream and complete accounting for all whey and lactose (by-products of cheese production). For lack of better information and given the need for a reasonable accounting procedure for estimating milk equivalents, the National Milk Producers Federation's estimates of MEs represents the best source available.

Table 2.5. Milk-equivalent conversion factors for imported dairy products: milkfat, solids-not-fat, and total solids basis

	Milkfat	Solids-not-fat	Total solids[a]
Quota cheese			
Blue mold	8.17	9.24	8.81
American cheddar	9.23	9.90	9.63
American other than cheddar	9.23	9.90	9.63
Edam/Gouda	7.90	8.93	8.52
Processed Edam/Gouda	6.70	8.93	8.04
Italian (IOL)	7.63	9.33	8.65
Italian (NIOL)	7.63	10.87	9.57
Swiss Emmenthaler	7.49	9.65	8.79
Swiss/processed Gruyère	8.17	9.66	9.06
Cheddar NSTLR	9.23	9.90	9.63
Other cheese NSPF	7.49	9.65	8.79
Cheese (low-fat)	0.11	19.77	11.91
Nonquota cheese			
Bryndza	7.22	8.43	7.95
Gammle/Nokkelost	7.22	8.43	7.95
Gjetost	7.90	10.00	9.16
Goya Cheese	7.90	10.00	9.16
Pecorino	8.17	9.30	8.85
Roquefort	8.77	8.98	8.90
Stilton	7.22	8.43	7.95
Other miscellaneous	7.90	10.00	9.16
Other dairy quota			
Butter	22.48	0.12	9.06
Butter oil	27.19	0.01	10.88
Butterfat mixtures	7.63	8.30	8.03
Ice cream (liter)	5.45	1.73	3.22
Milk and cream (liter)	11.99	0.58	5.14
Dried skim milk	0.22	11.10	6.75
Dry buttermilk/whey	1.36	10.60	6.91
Dry whole milk	7.22	8.19	7.80
Malted milk	13.62	5.77	8.91
Evaporated milk	2.15	2.05	2.09
Condensed milk	2.32	2.27	2.29
Chocolate cream (regular)	2.29	2.49	2.41
Chocolate cream (low-fat)	1.23	2.94	2.26
Dry cream	23.98	1.15	10.28
Feed (with solids)	0.30	11.07	6.76
Other (nonquota)			
Casein	0.08	10.76	6.49
Whey products	0.33	10.88	6.66
Lactose	0.00	11.51	6.90
Yogurt	0.44	3.03	1.99

Source: National Milk Producers Federation.

[a]Uses a 40/60 weight between the milkfat and solids-not-fat milk-equivalent conversion.

Table 2.6. Milk-equivalent conversion factors for commercial disappearance, exports, and government use of U.S. dairy products: milkfat, solids-not-fat, and total solids basis

	Milkfat	Solids-not-fat	Total solids[a]
Cheese			
Fresh cheese	9.23	9.90	9.63
Blue-veined cheese	8.17	9.66	9.06
Cheddar cheese	9.23	9.90	9.63
Processed cheese	7.44	9.30	8.55
Colby cheese	9.23	9.90	9.63
Grated/powdered cheese	7.63	10.87	9.57
All other	6.68	10.76	9.13
Other dairy products			
Butter	21.80	0.12	8.79
Butter oil	27.19	0.01	10.88
Casein	0.08	10.76	6.49
Lactose	0.00	0.00	0.00
Dry whey	1.36	10.60	6.90
Condensed/evaporated milk	2.15	2.05	2.09
Nonfat dry milk	0.22	11.58	7.04
Dry whole milk	7.22	8.19	7.80
Ice cream	3.65	0.99	2.05
Yogurt	0.44	3.03	1.99
Fluid milk	1.60	1.17	1.34
DEIP and government programs[b]			
Butter	21.80	0.12	8.79
Butter oil	27.19	0.01	10.88
Cheese	9.23	9.90	9.63
Nonfat dry milk	0.22	11.58	7.04
Dry whole milk	7.22	8.19	7.80

Source: National Milk Producers Federation.

[a]Uses a 40/60 weight between the milkfat and solids-not-fat milk-equivalent conversion.

[b]DEIP = Dairy Export Incentive Program.

Theory of Milk Marketing

The theory of milk marketing presented in this section will help the reader better understand how milk is priced and marketed in the United States within federal milk marketing orders. This is the only section of the book that requires some understanding of economics because a simple economic model is presented. A brief background on marketing conditions during the early decades of this century is introduced. Present-day dairy marketing policies are directly descended from efforts to correct the inequities of adverse marketing conditions during the Great Depression. The economic model offered in this section is based on the pioneering work of Gaumnitz and Reed in 1937, and later by Harris (1958) and Ippolito and Masson (1978).

Early Marketing Conditions

Our nation's cities at the turn of the century were small and were supplied milk by local dairy producers. These producers were referred to as "producer-distributors" since they both produced and distributed their own milk supplies. Producer-distributors often delivered their wares directly to the consumer's doorstep. This created a very close relationship between farmer and consumer. Then, as cities grew in size, it became necessary to secure fluid milk supplies from an increasingly distant market. It was no longer practical for farmers to market their milk themselves. As a result, professional milk distributors took over the business of purchasing milk from farmers, processing it, and distributing it to consumers. These distributors had tremendous investments in country milk-receiving stations, where farmers delivered their milk, and processing facilities located in the city that produced bottled milk and other dairy products.

Large proprietary milk distribution systems arose (1) from the need for economies of scale in the assembly and distribution of milk and (2) because large capital investments were required. Large distributors purchased milk from many small unorganized producers. These producers often felt they did not receive favorable prices and that they were in a poor bargaining position. In addition, when supplies of milk were plentiful, distributors often told some producers they no longer had a market for their milk. This would force these producers to find an alternative market for their milk at much lower prices. The small producers also could not verify the weights and tests distributors made on their milk. It was out of this environment that bargaining cooperatives emerged to provide dairy farmers better bargaining power with the large and powerful distributors. Thus began dairy cooperative marketing associations. These bargaining cooperatives grew rapidly following World War I. Gaumnitz and Reed (1937) list five functions that early bargaining cooperatives provided their members:

1. To act as agents of the producers in bargaining for prices with distributors.
2. To guarantee the producers a market for their milk.
3. To check the weights and tests of the producers' milk.
4. To ensure the payment to producers of sums owed by distributors.
5. To furnish market information.

The task of marketing a member's milk seemed simple. Cooperatives would agree to act as the sole sales agent and to market all of a member's milk, and would return the proceeds of such sales to the producer less the expense of operating the cooperative. But bargaining for milk prices with distributors was no easy task. Marketing cooperatives initially settled for flat pricing. Under this system, distributors paid cooperatives one flat price for fluid milk and milk intended for manufacturing uses. In effect, it was a weighted average of the higher fluid milk price and the lower manufacturing grade price. For a while this system worked. It began to crumble, however, since it did not take into consideration the fact that not all distributors sold the same proportion of their milk for fluid uses. Those distributors that sold a higher proportion of their milk in fluid form, and thus received a higher price for their sales, were at a distinct advantage over their competitors that processed a greater proportion of their milk into manufactured dairy products. To avoid a loss in revenue, some distributors cut off dairy producers during surplus times of the year in order to reduce purchases of milk intended for manufacturing purposes. Distributors also argued that prices had to fall given the apparent surplus of milk in the market during the spring flush. Thus producers and their cooperatives quickly realized that flat pricing was not a workable method to price milk. What was needed was an alternative method that would not only allow dairy farmers to sell all of their milk supplies but would also result in stable milk prices. This evolved into a pricing method called "classified pricing."

Unique Considerations for Milk and Milk Marketing

Before discussing the evolution and theory of classified pricing, it is important to review the unique properties of milk, how it is marketed, and the operation of early milk markets. The following factors must be taken into consideration when understanding how milk is priced and marketed.

1. The production of milk is seasonal. As described earlier, milk production peaks in the early spring with improved forage.
2. Milk is a perishable product. As such, it must be marketed within two days after being harvested from the cow. It also must be produced, distributed, and sold under strict sanitary conditions in order to protect consumers from milk-borne diseases. The perishability of milk drastically affected dairy farmers' bargaining position with milk distributors since farmers could not store raw milk and bargain for better prices later. This was particularly true of early bargaining cooperatives since many did not own processing facilities.
3. Milk is a bulky product. Transporting raw milk is very expensive due to the large amount of water in raw milk and the need to transport it under sanitary conditions.
4. The demand for fluid milk is perceived to be highly inelastic. That means consumers continue to purchase the same amount of milk regardless of changes in its price. It is generally recognized that a rise in fluid milk prices would usually result in very little loss in milk consumption. This has been used as a justification for passing on higher milk prices to consumers.
5. Fluid milk marketing can create a surplus problem. Because of the seasonal nature of milk production and because early markets could not transport milk from distant markets, fluid milk had to be produced in excess of local market needs in some months in order to have sufficient quantities available at other times. Thus, the surplus of fluid milk had to be manufactured into storable dairy products. Another reason for the surplus is that cows produce milk seven days a week whereas bottlers typically purchase milk less than five days a week.

Milk production and marketing up until the early 1950s centered exclusively around major cities in what was known as a "milkshed." This model of regional milk production is depicted in Figure 2.7 and consists of defined milk zones. The city center of the milkshed represented the market demand for both fluid milk and manufactured dairy products. Dairy producers who were willing to supply the city with fresh milk for bottling purposes located near the outskirts of the city due to the high cost of transportation outlined above. These producers located in the milk zone and specialized in fluid milk production. They met the strict sanitary standards set by the city milk inspectors. The next zone represented producers who provided fresh cream to the market. Fresh cream was separated from raw milk on the farm, which made it less expensive to transport to the city. The zone farthest from the city was the butter zone. Dairy producers located there to produce milk for manufacturing grade purposes. Processing plants that manufactured butter and nonfat dry milk, for example, were located in this zone. Such products could easily be marketed into the city center since their transportation costs were minimal relative to their economic value.

The price of milk in this model varied with the distance a producer was located from the city center. This is illustrated in Figure 2.8.

Figure 2.7. Relationship of supply zones in a milkshed

Producers nearest the city received the highest price since they were closest to the market and met the strict requirements of the city's sanitary regulations. Prices for fluid milk were determined by line P^m and fell the farther one moved from the center of the city. This accounted for the high cost of transportation required to move bulky sources of fluid milk. The greater the shortage of fluid milk in the city, the farther milk was pulled in from outside the milk zone. Differences in prices by location were justified by transportation costs.

Producers who marketed cream from the farm received lower prices than if they sold raw milk for fluid purposes and were located in the milk zone. The cream prices also fell the farther from the city center as seen on line P^c. Finally, producers located in the distant butter zone received the lowest price as seen on line P^b. Producers in both the cream and butter zones were not eligible to market fluid milk into the city center. At that time they didn't meet the strict sanitary requirements of producers in the milk zone and therefore received lower milk prices. But their production costs for milk were lower since their milk was produced under less sanitary conditions. For producers in the butter zone, this milk was manufactured into storable dairy products that were marketed over much greater distances than raw fluid milk.

The purpose of presenting this milk model is to illustrate how milk is priced in relation to city markets. On the one hand, most milk today is

Figure 2.8. Relationship between farm prices and distance from the market

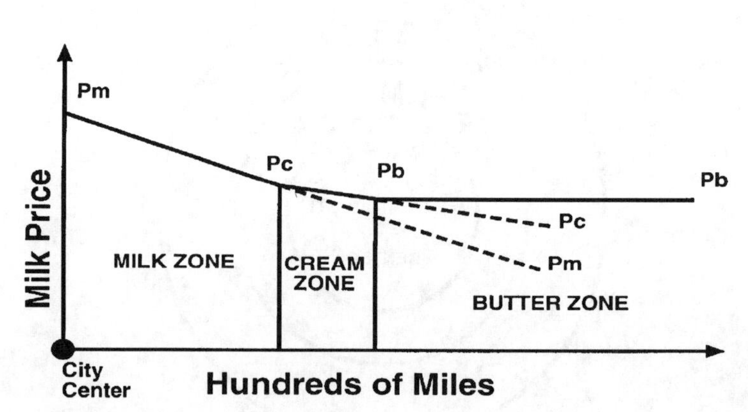

produced well beyond the city center and meets the highest sanitary condition. On the other hand, federal milk marketing orders are still centered around large cities and have pricing zones that lower producer prices the greater the distance from the cities. In addition, most fluid plants today are still located near city centers, and most manufacturing plants are located in rural areas. Thus some parts of the pricing model presented are still applicable today.

Theory of U.S. Milk Pricing

Bargaining cooperatives' difficulty enforcing flat pricing led to experiments with alternative pricing plans. During surplus periods, distributors argued for a lower flat price since an increasing percentage of their sales were for lower-valued manufactured products. Dairy cooperatives, on the other hand, needed to dispose of all of their members' milk for the highest prices possible. In order to accomplish this, the cooperatives bargained with distributors to purchase all of their members' milk at all times of the year. According to Gaumnitz and Reed (1937),

> It was proposed that the distributor show the producers exactly the quantity he sold for different uses, and that a basis of payment be arranged according to the quantities of milk sold in each of these classes. The plan is usually known as the classification plan and sometimes as the use plan.

Under classified pricing, bargaining cooperatives set up a schedule of prices distributors paid based on how milk was used for processing. Thus, those distributors that sold a higher percentage of their milk in fluid form paid a higher price to the cooperatives than those with a lower percentage of use for fluid purposes. This system was superior to the flat pricing system since distributors were put on a more equal footing. Bargaining cooperatives would then take the proceeds from the sale of their members' milk at the alternative prices and would distribute them in some equitable fashion to their members. This plan was first used on an extensive scale in Boston, Washington, D.C., and Philadelphia in about 1918 and spread in use thereafter.

The problem with the classified pricing plan back then was that it was entirely voluntary. When the Great Depression hit, consumers no longer had money to buy fluid milk products. As a result, surpluses mounted, and this voluntary program broke down. Farm prices plummeted, and violence erupted as farmers faced financial hardships. Marketing conditions were greatly disrupted. It was under these strenuous and unusual conditions that federal milk marketing orders were legislated. Federal milk marketing orders formalized the classified pricing system and enforced minimum prices for alternative classes of milk use in orders where dairy producers voted for them. Distributors' purchases were then regulated, they were required to pay dairy farmers at least the minimum class prices based on their use, they had to pool the results of their sales in a federal order pool, and they had to submit to having their records audited in order to guarantee producers were paid for the milk they delivered.

The application of classified pricing involves what economists refer to as "discriminatory pricing." Under discriminatory pricing, a seller establishes alternative prices for a product in order to maximize profits. In the case of milk, bargaining cooperatives in the past attempted to divert sufficient supplies of seasonal surpluses of milk away from a geographically isolated city market to lower-priced manufacturing markets in order to stabilize and enhance overall producer prices. This is illustrated in Figure 2.9 from a model developed by Ippolito and Masson (1978).

The demand for fluid milk products is denoted by the line D_I in the first panel called "the Class I market." The vertical axis represents milk prices, while the horizontal axis represents quantities of milk. Note that

Figure 2.9. Model of discriminatory pricing of milk in the United States

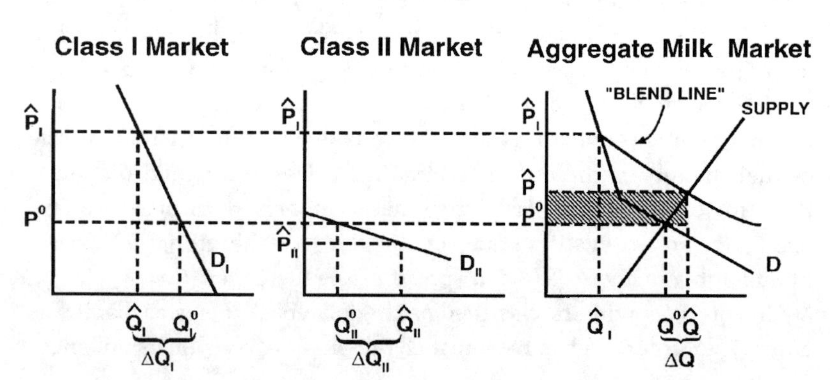

this demand curve is very steep (what economists call "inelastic"). The quantity of milk used for fluid purposes is therefore unresponsive to changes in milk prices. The second panel presents the demand curve for all manufactured dairy products, or the Class II market, and is denoted D_{II}. This schedule is less steep and is therefore more responsive to changes in the price of milk. The third panel shows the combined demand schedules for both Class I and Class II milk (labeled D) and the aggregate supply curve for all dairy producers.

Under an unregulated market, the intersection of the supply and demand curve D yields an equilibrium milk price of P^0 that is faced by both milk handlers and dairy producers. This in turn results in Q_I^0 demanded for fluid uses (in panel 1) and Q_{II}^0 demanded for manufacturing uses (in panel 2). The imposition of a regulated market that enforces discriminatory pricing results in an alternative equilibrium. Regulators will require milk handlers to pay a higher price for milk (the Class I price) used for fluid purposes. This can be seen as $\hat{P}_I$ in panel 1. The demand for fluid milk will be satisfied first since its market price is higher than that for other uses. Note that a higher price for fluid milk will lower the consumption of milk for fluid purposes from Q_I^0 to $\hat{Q}_I$. At the same time, milk not used for fluid uses will be diverted to Class II purposes by milk handlers offering a much lower price, the Class II price ($\hat{P}_{II}$), for milk used for manufacturing purposes. Class II milk therefore becomes a residual demand for milk in excess of fluid uses. This lower price expands consumption of milk for manufacturing uses from Q_{II}^0 to $\hat{Q}_{II}$. Since the demand curve for Class II milk is much more responsive to changes in milk prices, it is therefore able to absorb the additional milk supplies diverted from Class I purposes.

Under classified pricing, dairy farmers receive a blend of both Class I and Class II prices that may be characterized in the following average revenue (AR) schedule:

$$AR = (P_I Q_I + P_{II} Q_{II})/(Q_I + Q_{II})$$

The average revenue schedule varies with classified prices and the elasticity of demand for each demand schedule. Thus the average revenue schedule reflects the supply-inducing price that dairy farmers face under classified pricing. New equilibrium prices and quantities for a regulated market are determined in panel 3 by the intersection of AR (depicted as the blend line) and the supply schedule. Note that discriminatory pricing therefore results in a higher equilibrium market price facing producers: $\hat{P}$. In addition, the equilibrium quantity of milk supplied to the market increased from Q^0 to $\hat{Q}$. This results in an increase in total revenue, which is indicated by the shaded region in panel 3. Discriminatory pricing via classified prices thus results in a higher total revenue to milk producers, but also results in overproduction as evidenced by $\hat{Q}$–Q^0.

REFERENCES

Campbell, John R. and Robert T. Marshall. 1975. *The Science of Providing Milk for Man.* New York: McGraw-Hill Publishing Co.

Cochrane, Willard W. and C. Ford Runge. 1992. *Reforming Farm Policy: Toward a National Agenda.* Ames: Iowa State University Press.

Gaumnitz, E.W. and O.M. Reed. 1937. *Some Problems Involved in Establishing Milk Prices.* DM-2 Marketing Information Series. Agricultural Adjustment Administration, U.S. Department of Agriculture. Washington, D.C.: U.S. Government Printing Office, September.

Haidacher, Richard C. 1992. "Market Demand Structure for Dairy Products in the United States." In *Market Demand for Dairy Products,* edited by S.R. Johnson, D. Peter Stonehouse, and Zuhair A. Hassan. Ames: Iowa State University Press.

Harris, Edmond S. 1958. *Classified Pricing of Milk: Some Theoretical Aspects.* Technical Bulletin 1184. Agricultural Marketing Service, U.S. Department of Agriculture, Washington, D.C., April.

Ippolito, Richard A. and Robert T. Masson. 1978. "The Social Cost of Government Regulation of Milk." *Journal of Law and Economics,* vol. 21, April.

Jacobson, Robert. 1992. *Calculating Milk Equivalents: Milkfat or Total Solids Basis.* P-2, Dairy Markets and Policy Issues and Options. Cornell University, August.

National Milk Producers Federation. 1995. "Trends in U.S. Dairy Demand." In *Dairy Market Report.* Produced for Dairy Management, Inc., by the National Milk Producers Federation, February.

U.S. Department of Agriculture. Agricultural Research Service. 1976. *Composition of Foods: Dairy and Egg Products, Raw, Processed, Prepared.* Agriculture Handbook 8-1. Principal investigators: Linda Posati and Martha Louise Orr. Washington, D.C.: Consumer and Food Economics Institute, November.

U.S. Department of Agriculture. Agricultural Stabilization and Conservation Service. Commodity Analysis Division. 1990. *Methodology of Calculating the Milk Equivalent Total Solids Basis of CCC Purchases of Surplus Dairy Products and of Imports of Dairy Products: As Required by the Food, Agriculture, Conservation, and Trade Act of 1990.* December 19.

U.S. Department of Agriculture. Economic Research Service. 1995. *Dairy Yearbook.* Statistical Bulletin 924. December.

U.S. Department of Agriculture. National Agricultural Statistics Service. 1994. *Milk Production, Disposition and Income: 1993 Summary.* Da 1-2(94). May.

Chapter 3

Fluid Milk Products

From the time that Sir Thomas brought 100 cows from England to the new settlement Jamestown in 1611 until the early part of the nineteenth century, milk was consumed on or very near the farm. Most families had one or two cows for producing milk, cheese, and butter and were willing to sell any surplus to neighbors or friends. Then, as cities began to grow and demand for milk increased, so did the dairy industry. Many farms sprang up within U.S. cities. Milk dealers began to buy milk from local farmers by the early 1800s and sold it from the back of a horse and cart. By the mid-1800s, as city land became scarce due to population pressures and sanitary concerns, farmers and their cows began to move outside the city limits. Distillery owners, who replaced some of these local dairy farmers, found it profitable to feed distillery slop to dairy cows. They produced and marketed under highly unsanitary conditions an inferior milk product called "swill milk." This milk was fed to many poor children in the nation's cities and was linked to high infant mortality rates (Selitzer 1976). In addition, milk was adulterated with starch, flour, plaster of Paris, chalk, and water (Giblin 1986). Thus began the campaign for pure milk that started during the bleakest period in the history of the U.S. dairy industry. Swill dairies were shut down, sanitation and adulteration laws were passed, and the emerging railroad industry in the 1840s brought quality supplies of fresh milk from the countryside to the cities.

Today Americans often take for granted the available supplies of pure wholesome milk in local supermarkets. Advances in animal husbandry, transportation, and processing and distribution as well as strict sanitary standards were required to bring forth the products American consumers enjoy today. This chapter focuses on the processing of and demand for fluid milk products.

Fluid Milk Products and Consumption

The fluid milk industry represents big business in the United States. Total retail sales for fluid milk products in 1994 were $28.4 billion (Milk Industry Foundation 1995). Those sales were based on 55 billion pounds of fluid milk products and 3.3 billion pounds of cream and specialty dairy products such as sour cream and dips, eggnog, and yogurt. Almost all fluid milk sales in 1994 were from supermarkets and dairy/convenience stores. Other outlets were, for example, the military and schools. Of all fluid milk sold in 1994, three-quarters was sold in plastic containers and the balance in paper containers. The most popular size container in 1994 was the gallon jug (64 percent), followed by the half-gallon (19 percent), half-pint (9 percent), quart (4 percent), bulk (2 percent), and pint and other sizes (2 percent).

The USDA lumps a number of products into the broad category of fluid milk and cream products. On the beverage side this consists of both plain and flavored milk such as whole milk, ½ percent, 1 and 2 percent milk, skim milk, buttermilk, and eggnog. The standards of identification for fluid milk products under the *Code of Federal Regulation* (*CFR*) require a minimum 3.25 percent milkfat for whole milk, 0.5 to 2.0 percent milkfat for low-fat milk products, and less than 0.5 percent for skim milk products.[1] All of these fluid milk products must also have at least 8.25 percent milk solids-not-fat. Standards of identity for cream include half-and-half, light and heavy cream, light whipping cream, and sour cream.

The International Dairy Foods Association announced that the Food and Drug Administration (FDA) published a final rule on November 20, 1996, to eliminate the federal standards of identity for twelve lower-fat dairy products including low-fat milk, skim (nonfat) milk, and lower-fat versions of all cultured products (except yogurt) (USDHHS 1996). All lower-fat fluid milks and cultured products (except yogurt) are now subject to the nomenclature that stems from the passage of the Nutrition Labeling and Education Act of 1990. The standards of identity for whole milk, cottage cheese, sour cream, and other parent standards remains unaffected by the final rule.

The lower-fat versions of fluid milk are now subject to the requirements of the FDA's "general standard," which permits foods to be named by using a defined nutrient claim such as "reduced fat," "light," "nonfat," and "fat free" in conjunction with standardized terms like milk and cottage cheese. Under the new system, lower-fat products must (1) meet nutrient content descriptor definitions for total fat content, (2) be nutritionally equivalent to the reference (i.e., "full-fat") standard of identity, and (3) meet all other provisions of the reference standard. These changes will be implemented on January 1, 1998. Examples of the changes in labeling requirements for low-fat fluid milk products are contained in Table 3.1

Fluid milk products have become diversified over the years to meet the needs and preferences of consumers. Selitzer cites Americans' concern about high-fat foods—particularly the fat content of whole milk—beginning just after World War II. An article in the October 1950 issue of

Table 3.1. Revised labeling requirements for select lower-fat fluid milk products, 1998

Current		New	
Product name	Standard	Product name	Standard
Low-fat milk, 2% milkfat	2% milkfat by weight	Reduced-fat milk or reduced-fat milk, 2% milkfat	Minimum of 25% reduction in total fat per reference amount[a]
Low-fat milk, 1.5% milkfat	1.5% or 1½% milkfat by weight	Reduced-fat milk or reduced-fat milk, 1.5% milkfat	Minimum of 25% reduction in total fat per reference amount[a]
Low-fat milk, 1 % milkfat	1% milkfat by weight	Low-fat milk or low-fat milk, 1% milkfat	Maximum of 3 g total fat per reference amount[a]
Skim or nonfat milk	Less than 0.5% milkfat by weight	Skim, nonfat, or fat-free milk	Less than 0.5 g total fat per reference amount[a]

[a]Reference amount for all fluid milks is 240 ml (1 cup or 8 fl oz.) The percent reduction is relative to whole milk.

Source: International Dairy Foods Assocation 1996.

Note: "Light" will be permitted in the name of milk if the fat content is reduced 50% (i.e., for products currently called 1.5%, 1%, and ½%).

the *Saturday Evening Post* described research by Dr. Gofman of the University of California in which he associated arteriosclerosis and heart disease with the consumption of dairy foods. The article concerned the public, and demand for lower-fat fluid milk products skyrocketed. As a result, whole milk consumption dropped from 290 pounds per person in 1950 to around 76 pounds in 1994. At the same time, demand for low-fat fluid products increased. Table 3.2 indicates that per capita fluid milk consumption of 1 percent, 2 percent, and skim milk increased significantly from 1970 to 1994. A small percentage of this increase was in the form of flavored milk, which is popular in the school lunch program.

Table 3.2. Per capita consumption in pounds of fluid milk, 1970–94

	Plain fluid milk						Flavored milk and milk drink			Total fluid milk
		Low-fat								
Year	Whole	2%	1%	Total	Skim	Total	Whole	Low-fat	Total	
1970	213.5	28.0	1.8	29.8	11.6	255.0	5.6	3.0	8.6	269.1
1971	208.7	30.9	3.0	34.0	12.3	255.0	6.2	2.6	8.8	269.4
1972	200.4	34.6	4.6	39.2	12.4	252.0	7.1	2.5	9.6	267.1
1973	190.4	39.1	4.0	43.1	13.8	247.3	7.3	2.7	10.0	262.3
1974	180.0	38.2	7.6	45.8	13.9	239.7	6.7	2.6	9.4	253.7
1975	174.9	40.5	12.7	53.2	11.5	239.6	6.3	3.3	9.7	254.0
1976	168.4	43.9	13.2	57.1	11.6	237.1	6.8	4.0	10.8	252.6
1977	160.7	47.4	13.7	61.1	11.9	233.7	6.6	4.8	11.4	249.7
1978	154.9	49.6	14.6	64.2	11.5	230.5	6.1	4.9	11.1	246.0
1979	149.3	52.4	14.6	67.0	11.6	227.8	5.5	5.0	10.5	242.6
1980	141.7	54.7	15.3	70.1	11.6	223.3	4.7	5.3	10.0	237.4
1981	136.3	57.0	15.6	72.6	11.3	220.2	3.7	5.6	9.3	233.5
1982	130.3	58.3	15.3	73.5	10.6	214.4	3.1	5.5	8.6	227.1
1983	127.1	60.7	14.8	75.4	10.6	213.1	3.2	5.9	9.1	226.5
1984	123.0	64.2	14.3	78.5	11.6	213.1	3.8	6.0	9.8	227.3
1985	119.7	68.5	14.7	83.3	12.6	215.8	3.7	6.0	9.7	229.7
1986	112.9	71.8	16.3	88.1	13.5	214.5	3.5	6.3	9.9	228.6
1987	108.5	74.0	15.6	89.7	14.0	212.2	3.4	6.6	10.1	226.5
1988	102.4	74.6	15.3	89.9	16.1	208.4	3.3	6.6	9.9	222.3
1989	94.4	79.1	17.2	96.3	20.2	210.9	3.1	6.5	9.6	224.2
1990	87.6	78.4	19.9	98.3	22.9	208.7	2.8	6.6	9.4	221.7
1991	84.6	78.9	20.8	99.7	23.9	208.2	2.7	6.8	9.5	221.2
1992	81.4	78.5	21.0	99.4	25.0	205.9	2.7	6.9	9.6	218.6
1993	77.8	76.7	20.5	97.1	26.7	201.7	2.7	6.9	9.6	214.3
1994	75.8	74.9	20.7	95.6	28.8	200.2	2.7	7.1	9.8	213.0

Source: Putnam and Allshouse 1996.

The overall demand for fluid milk products declined from 269 pounds per person in 1970 to 213 pounds in 1994.

While there was a decrease in consumption of high-fat fluid milk, just the opposite was true for cream and sour cream products. Table 3.3 indicates that per capita consumption of half-and-half (a half-cream and

Table 3.3. Per capita consumption in pounds of other fluid milk products, 1970–94

	Cream						
Year	Half-and-half	Light	Heavy	Total	Sour cream	Total cream and sour cream	Eggnog
1970	2.9	0.4	0.5	3.8	1.1	4.9	0.3
1971	2.7	0.3	0.5	3.6	1.2	4.8	0.4
1972	2.6	0.3	0.5	3.4	1.3	4.7	0.5
1973	2.6	0.4	0.6	3.6	1.3	4.9	0.4
1974	2.4	0.4	0.5	3.4	1.5	4.8	0.4
1975	2.4	0.4	0.6	3.3	1.6	5.0	0.4
1976	2.4	0.3	0.6	3.4	1.6	5.0	0.4
1977	2.4	0.3	0.6	3.3	1.7	5.0	0.4
1978	2.4	0.3	0.6	3.3	1.7	5.0	0.4
1979	2.4	0.3	0.6	3.3	1.8	5.1	0.4
1980	2.4	0.2	0.7	3.4	1.8	5.2	0.4
1981	2.5	0.2	0.7	3.4	1.8	5.3	0.4
1982	2.5	0.3	0.7	3.5	1.9	5.4	0.4
1983	2.6	0.3	0.8	3.7	2.1	5.8	0.5
1984	2.8	0.3	0.9	4.0	2.2	6.3	0.5
1985	3.0	0.4	1.0	4.4	2.3	6.7	0.5
1986	3.2	0.4	1.1	4.7	2.4	7.0	0.5
1987	3.1	0.4	1.1	4.7	2.4	7.1	0.5
1988	3.0	0.4	1.2	4.6	2.5	7.1	0.5
1989	3.1	0.4	1.3	4.8	2.5	7.3	0.5
1990	3.0	0.3	1.3	4.6	2.5	7.1	0.5
1991	3.1	0.3	1.3	4.6	2.6	7.3	0.4
1992	3.2	0.3	1.3	4.8	2.7	7.5	0.5
1993	3.2	0.4	1.4	4.9	2.7	7.6	0.4
1994	3.1	0.3	1.4	4.9	2.7	7.6	0.4

Source: Putnam and Allshouse 1996.

Note: Totals have been rounded off.

half-milk product used in coffee) increased from 2.9 pounds per person in 1970 to 3.1 in 1994. Also, consumption of heavy cream increased from half a pound per person in 1970 to 1.4 pounds in 1994. But sour cream consumption increased the most, from 1.1 pounds in 1970 to 2.7 pounds in 1994. When it comes to coffee, pies, and baked potatoes, American consumers, it seems, are unwilling to give up on taste!

Receiving and Processing Fluid Milk Products

Milk undergoes careful processing under strict sanitary conditions before it can be marketed as fluid milk. It must first be transported under sanitary conditions from the dairy farm to the processing plant in a timely manner. Processing involves standardizing the amount of milkfat in the fluid product, pasteurizing milk in order to ensure that it is safe to drink, and then homogenizing milk by breaking up the suspended fat particles in order to ensure that the milkfat won't be separated as cream before the consumer drinks it.

Immediately after milking, farmers store milk in refrigerated stainless steel bulk tanks located in the milk house. The purpose of bulk tanks is to store milk in a sanitary environment, to lower the temperature of the milk after milking,[2] and to prevent the milk from separating. Some dairy producers have a plate chiller that reduces the temperature of the milk before it enters the bulk tank. Plate chillers are usually more efficient than bulk tanks in lowering the temperature. Once in the bulk tank, milk is agitated periodically in order to prevent separation of the cream and to help keep the milk cool. The milk is then picked up by the milk hauler either every day or every other day depending on the capacity of the bulk tank and arrangements made with the milk hauler.

The milk hauler represents an important link in the chain of milk quality. The hauler must examine the milk and determine if its temperature, appearance, flavor, and odor are acceptable (Campbell and Marshall 1975). The hauler must also measure the volume of milk in the bulk tank and take samples of milk before loading it into the milk truck. (These samples are tested later at the fluid processing plant before the milk is unloaded into the plant's milk silo.) After the farm samples are drawn and the milk passes the hauler's initial inspection, it is loaded onto insulated milk trucks. Milk must be chilled on the farm before the transfer since the milk truck is not refrigerated. There are two types of milk trucks. The smaller trucks, called "bob trucks" or "straight trucks," have a capacity of between 17,000 and 30,000 pounds of milk and essentially collect milk from several producers along a defined milk route. The milk is then either hauled directly to a processing plant or it is taken to a receiving station where it is combined with more milk, refrigerated, and loaded onto a much larger truck. More and more, farm milk is being picked up by larger insulated trailers with a capacity of up to 50,000 pounds. These farm pickups go directly to a designated plant. A new trend on larger dairy farms is to offload the milk directly from the plate chiller to the tanker truck, thereby bypassing the capital investments needed for a refrigerated bulk tank or large insulated upright silos. Only large dairy farms with enough capacity to fill a tanker truck once or more a day have the ability to do this.

Milk is then transported to the fluid processing plant, where it arrives at the receiving department, which consists of facilities for unloading, washing, and sanitizing milk tankers (Henderson 1971). The individual samples taken by the milk hauler are tested for antibiotics, bacteria count, and general milk quality. Some samples are also spot-checked for aflatoxin. These sample tests not only determine if the milk is acceptable for processing but also gauge the composition and quality of the milk. In fact, many producers are paid on the basis of these tests. For example, some producers are paid on the basis of butterfat content, solids-not-fat or protein content, and somatic cell and bacteria count for individual loads of milk. If all of the samples meet the quality standards set by the Pasteurized Milk Ordinance (PMO), the milk is loaded into large insulated milk silos and commingled with the plant's milk supply. (The PMO is decribed more fully in Chapter 9.) The PMO is a national standard for the sanitary control of Grade A milk and dairy products. If any individual test fails to meet the strict requirements of the PMO, then the whole load is rejected, and the milk is dumped or disposed of for animal feed. These tests are done before the milk is commingled with the plant's milk supply in order to prevent dumping a large silo of milk.

Before the milk gets inside the milk silo, it may be further chilled if necessary and is often clarified using a cold milk clarifier. Clarifiers use centrifugal force to remove insoluble suspended materials from the milk. They essentially operate as milk filters. Once in the silo, the milk is agitated periodically and the temperature recorded.

Milk is often processed and distributed to retail stores less than 24 hours after it leaves the farm. Milk must undergo rigorous processing under strict conditions outlined in the PMO before it can be sold in a retail store. First, milk flows from the silos to a separator where the cream is separated from skim milk. Milk is often heated to 90° to 100°F (32° to 38°C) before it is separated. Prior to the invention of separators, milk was stored in a special tank for 48 hours where the cream was allowed to naturally rise to the top before it was hand separated. (Cream is basically concentrated milkfat.) But this process took days, and spoilage often occurred. De Laval invented a hand-cranked separator just before the turn of the century. Unlike hand-cranked separators, modern cream separators can do their job within seconds. After separation, part of the cream and all of the skim milk is then recombined in the proper proportion to produce the desired product (i.e., skim milk, 1 percent, 2 percent, or whole milk). This process often results in excess cream supplies since even whole milk, which has a minimum milkfat content of 3.25 percent, falls below the natural fat content of raw cow's milk during most months of the year.

After milk is separated, it is heated and homogenized in order to suspend the fat globules in milk and thereby prevent separation of cream. Milk homogenizers use piston-type pumps to force milk through a restricted valve under tremendous pressures to disrupt and break down the fat globules (Henderson 1971). These smaller fat globules remain suspended in the milk and won't rise to the top as will larger fat globules in raw milk.

Milk is then pasteurized after homogenization. Pasteurization was named after the French chemist Louis Pasteur, who discovered that bacteria were responsible for allowing beer, wine, and milk to turn sour and spoil (Giblin 1986). He discovered during experiments from 1860 to 1864 that heating wine for a few minutes between 140° and 158°F (60°–70°C) would prevent it from spoiling but would not impair its delicate flavor. Scientists soon realized that Pasteur's process also applied to milk and could destroy dangerous microorganisms that could produce and spread disease. In 1887, Austrian chemist F. Soxhlet adopted Pasteur's heat method and developed the first homemade device to sterilize milk in nursing bottles. Pasteurization then became widespread to prevent the spread of tuberculosis and brucellosis and to contain outbreaks of such diseases as typhoid fever, diphtheria, and scarlet fever. Before the

International Medical Congress in 1881, Dr. Ernest Hart reported the results of a study that implicated infected milk as responsible for 50 epidemics of typhoid fever, 15 outbreaks of scarlet fever, and 4 of diphtheria (Selitzer 1976). Widespread use of pasteurization first occurred in New York City when Dr. Henry Koplik opened the first milk depot in 1889 designed to pasteurize milk for infants. Later, pasteurization was championed by the great humanitarian Nathan Straus, then head of R.H. Macy's in New York City. Straus also opened a pasteurized milk depot on the East Third Street pier in a crowded neighborhood (Giblin 1986). These milk depots were a welcome alternative for mothers who were concerned about the dangers of swill milk. Later, after the turn of the century, city ordinances began to be passed, based on Straus's efforts, that required pasteurization before milk could be sold.

Modern pasteurization destroys undesirable microorganisms by heating milk to a high temperature for a short period of time. Low-temperature longtime pasteurization requires milk to be heated for 30 minutes at 145°F (63°C), whereas high-temperature short-time pasteurization requires just 15 seconds at 161°F (72°C) (Campbell and Marshall 1975).

Once pasteurized, milk is cooled and then fortified with vitamins A and D. Public health officials recommend milk be fortified with these essential vitamins to ensure that children receive adequate amounts. While milk naturally contains vitamin A and a small amount of vitamin D, the level of vitamin D varies due to breed of cow and production conditions (Henderson 1971). Prior to fortification, many city-dwelling children suffered from rickets caused by low levels of vitamin D in the diet and little exposure to the sun. Adding vitamin A or D is optional in whole milk but, if added, must be so labeled on the milk container. Low-fat and skim milk must be fortified to replace the naturally occurring vitamin A that is lost during the removal of the milkfat. If fortified, vitamin D must be at a concentration of 400 international units per quart, whereas vitamin A must be at a level of 2,000 international units per quart.

(In addition to vitamins, various syrups, such as chocolate and strawberry, are used in flavored milk products.) Milk is then packaged either in polyethylene jugs or polyethylene-laminated paper cartons. After the containers are sealed and stacked in cases, corrugated boxes, or carts, they are moved to a large refrigerated room where the milk is stored until loaded for delivery to a retail outlet.

Creams and Cultured Dairy Products

Creams and cultured dairy products make up a small portion of total milk sales. Fresh creams used to be a staple in American households, particularly when bottled milk was home-delivered. Cream was allowed to rise to the top in glass bottles and was then scooped out with a cream spoon and used for whipping cream, in coffee and breakfast cereals, and for cooking purposes. As home-delivered milk was replaced with store-bought homogenized milk, creams became more readily available as packaged products in the grocery store.

There are basically two types of cream available on the market today: (1) table cream, which is mostly mixed with coffee or used in cereal, and (2) whipping cream. Table cream or light cream contains 18 to 30 percent milkfat and has desirable properties when mixed with hot coffee (e.g., it won't feather). It is either pasteurized or ultrapasteurized from Grade A milk, but it may or may not be homogenized. Ultrapasteurization is where a product is heated to 280°F (137.8°C) or above and held for at least two seconds (Campbell and Marshall 1975). Half-and-half is a relatively new dairy product that was introduced when homogenization became popular. Consisting of half cream and half milk, half-and-half is a low-fat table cream that has a milkfat content of 10.5 to 18 percent. It is pasteurized or ultrapasteurized and may be homogenized. Heavy whipping cream has a higher milkfat content of at least 36 percent and is pasteurized or ultrapasteurized and may be homogenized. The *CFR* requires a minimum of 36 percent milkfat in heavy cream. Light whipping cream is cream that contains 30 to 36 percent milkfat. It is pasteurized or ultrapasteurized and may also be homogenized. Henderson (1971) notes that there has been a rapid decline in the sales of cream products due to (1) the development of half-and-half, (2) the effort of nutritionists to reduce fat in the American diet, and (3) the development of cream substitutes such as coffee whiteners, nondairy toppings, whipped cream substitutes, and aerosol creams.

Cultured dairy products also make up a small percentage of milk sales. Cultured dairy products are produced when milk is soured due to bacteria that produce lactic acid. This lowers the pH of the products and prevents the growth of other bacteria that produce disease and cause spoilage (Campbell and Marshall 1975). The term "cultured" is used since commercial manufacturers use pure bacterial cultures rather than relying on the unpredictable properties of bacteria that may exist naturally in the environment. Today, the major cultured dairy products that are commercially sold are buttermilk, sour cream and dips, and yogurt and cottage cheese (discussed in the next chapter).

Historically, buttermilk was the by-product of butter production. Essentially, it was the liquid material that was left after butter was churned. As far back as the colonial period, fresh buttermilk was considered a delicacy. Today, however, buttermilk is dried and used in the baking industry. Buttermilk today is drunk as a cultured product. Cultured buttermilk is pasteurized skim, low-fat, or whole milk that is soured by lactic acid–producing cultures, and it has the same milkfat and milk solids content of noncultured milk.

Sour cream products on the market today are sour cream, sour half-and-half, acidified sour cream, and acidified sour half-and-half. The *CFR* describes sour cream as resulting from the souring of pasteurized cream by lactic acid–producing bacteria. Also according to the *CFR*, sour half-and-half results from the souring of pasteurized half-and-half by lactic acid–producing bacteria. Sour cream contains no less than 18 percent milkfat, and sour half-and-half contains 10.5 to 18 percent milkfat.

Other fluid milk products available in many grocery stores include acidophilus milk and eggnog. Acidophilus milk was named after the bacterium *Lactobacillus acidophilus,* which is used as a culture in the preparation of this product. It is promoted as a health food and is prescribed by some physicians to control intestinal disorders. Eggnog is a seasonal beverage that is sold predominantly from Thanksgiving until New Year's Day. Eggnog is made from cream, milk, partially skimmed milk, or skim milk; egg yolks or egg yolk–flavored mix; and nutritive carbohydrate sweeteners. It contains not less than 6 percent milkfat and not less than 8.25 percent milk solids-not-fat.

Marketing and Distribution of Fluid Milk Products

The marketing and distribution of fluid milk products has undergone rapid change due to industrialization and economies of scale. Manchester (1983) notes that a major influence on the number and size of processors was the rapid shift in the cost of processing milk over time. When milk marketing was taken over by milk dealers in the early 1800s, equipment needs were very simple: a horse and wagon, milk cans, and a dipper. With the introduction of the glass milk container in 1878 came the introduction of the fluid milk factory and higher production costs (Selitzer 1976). Those production costs escalated further with the adoption of pasteurization after the turn of the century and then with the introduction of the paper carton in the late 1930s and 1940s. Tremendous mechanization and an increase in the size of the fluid plant occurred after World War II. Dairies that specialized in home-delivered milk and mom-and-pop stores that sold locally produced dairy products began to disappear and were replaced by large regional or national supermarket chains that promoted self-service merchandising. These chains often bought dairy products in large volumes and purchased prepackaged milk from large processors. As each new innovation or change in sanitary requirements was introduced, small processors with high fixed costs that could not make the rapid change were forced out of business. As a result, many small bottling plants that dotted most American towns prior to World War II began to disappear.

Lough (1991) reviews the dynamic structure of the fluid milk industry and outlines four major stages:

1. Growth of cities: before 1886
2. Formation of commercial industry: 1886–1920
3. Industry growth to maturity: 1920–62
4. New organization: 1962–88

Stages one and two are reviewed in Chapter 2 of this book. The growth of cities in stage one created the market for milk dealers when milk production moved outside the cities and farmers specialized in milk production. Stage two witnessed the introduction of sanitary requirements and technological innovations that created the factory system of fluid milk processing. Stage three represents the industry growth between the 1920s and early 1960s. Large-scale organization in the fluid milk business began to appear with the rash of mergers that formed major regional and national fluid milk companies. Well-known ones include Beatrice, Borden, Carnation, Fairmont, Foremost, Kraftco, and Pet. Lough notes that the eight largest dairy companies acquired 2,307 other dairy companies over the period 1920–61. While the number of fluid milk plants increased through the latter 1930s, plant consolidations and plant exits rapidly downsized plant numbers after World War II (Figure 3.1).

Tremendous changes during the third stage contributed to the reduction in the number of fluid milk plants. Local fluid markets gave way to regional markets as the transportation system improved. There was a rapid shift away from home-delivered milk to store sales of milk. This shifted market power away from the processor to the retail grocery store. State marketing orders and local sanitary regulations that limited interstate milk shipments were challenged in court. And a number of processing innovations were made that lowered per unit processing costs. As a result, by the end of stage three, both marketing regions and plants were much larger.

The industry further matured during stage four, during which the number of processing plants continued to decline and the remaining plants grew larger. The number of fluid processing plants fell from 4,103 in 1964 to just 640 in 1988 (Table 3.4). There were three major reasons for these plant closings during this period: (1) many smaller plants closed when the first-generation owner-manager retired, (2) a major account was lost (this was usually suffered by a medium-size firm), or (3) the company was bought out or merged with another company.

There was also a major shift in ownership during the fourth stage as the traditional seven large dairy companies that originated in stage two

Figure 3.1 Number of fluid milk bottling plants in the United States.

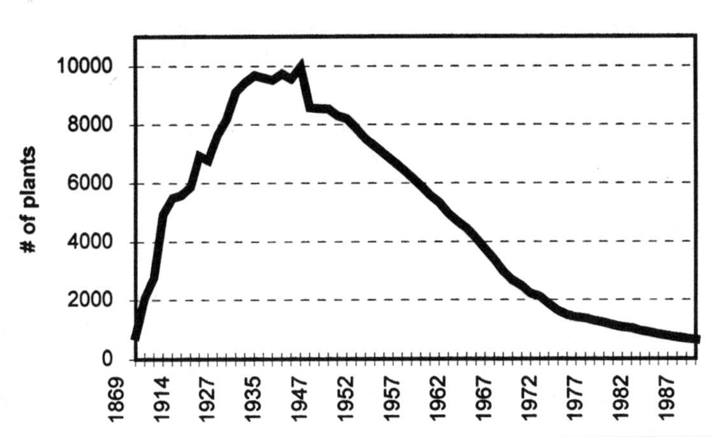

Table 3.4. Number of fluid milk-processing plants, by type of firm, in the United States

	1964	1970	1980	1988
National firms	264	196	159	102
Regional firms	71	57	14	40
Local firms				
Multiunit	229	118	49	44
Single-unit	3,234	1,609	667	368
Regional cooperatives	—	—	48	39
Local cooperatives				
Multiunit	115	102	37	32
Single-unit	152	83	27	15
Integrated supermarkets	38	51	65	—
Total	4,103	2,216	1,066	640

Source: Lough 1991.

gave way to vertical integration by large grocery store chains, national dairy cooperatives, and new proprietary owners. Federal Trade Commission actions in the 1960s prevented the top seven national dairy processors from making additional dairy acquisitions. But this didn't stop these companies from acquiring nondairy firms. As a result, the nondairy business of some of these companies dominated the dairy portion. Lough notes, "Foremost filed for bankruptcy, Borden bought Beatrice Dairy Products, Kraftco and Fairmont closed or sold the last of their fluid plants and Pet was sold." More recently, Borden was purchased by the investment firm of Kohlberg, Kravis, Roberts, and Company (Prepared Foods 1995).

The vertical integration of the fluid milk industry with large grocery store chains represents a major change in the dairy industry in the fourth stage. These large integrated supermarkets accounted for only 3.3 percent of total sales in 1964, but 18.4 percent by 1988 (Lough 1991). Many chains either owned fluid processing plants outright or developed centralized milk programs that contracted for milk on a regional or divisional basis. Dairy cooperatives also became integrated into the fluid milk business. Cooperatives have in some cases purchased fluid processing plants outright in order to find a market for their members' milk.

In other cases they purchased an interest in a fluid plant and hired outside management to run it. Mid-America Dairymen, Inc. of Springfield, Missouri, is an example of a dairy cooperative that has purchased these privately owned fluid milk processing plants to provide a financially stronger marketing opportunity for its dairy farm cooperative members.

Advertising and Promotion of Fluid Milk Products

Dairy farmers have been active in research, education, advertising, and promotion programs to increase the demand for dairy products for over 70 years. Most dairy farmers understand the need to educate the public on the nutritional value of dairy products and to promote their product. For most years, contributions by dairy producers for such efforts was voluntary. But the need for advertising and promotion became even more important in the 1970s and 1980s when fluid milk consumption began to decline, soft drink consumption skyrocketed, and dairy surpluses burgeoned.

Tauer and Forker (1987) trace the origins of advertising and promotion programs in the United States. Beginning in 1915, leaders of producer and dealer groups formed the National Dairy Council (NDC) to emphasize the importance of milk and dairy products in a healthy diet. Today the NDC administers grants and contracts to researchers in the nation's leading scientific, medical, and dental institutions to aid the discovery of new information about dairy foods. The NDC also develops programs that promote healthy eating habits. The American Dairy Association (ADA) was chartered in 1940 to advertise and merchandise dairy products. The ADA has over the years developed national programs in advertising, merchandising, public relations, and product and market research for milk and dairy products. The objectives of the ADA today are twofold: (1) to develop and implement programs that promote demand for U.S. dairy products and (2) to service the needs of member organizations as they relate to marketing, planning, and programming.

Both the NDC and ADA depended on voluntary support from dairy farmers to fund their activities. Leaders in both organizations became concerned in the late 1960s that activities in the organizations were overlapping and communications between the organizations were poor. As a result, the United Dairy Industry Association (UDIA) was formed in 1971 to eliminate duplicate efforts in promotion, education, and research. The UDIA was organized to serve as an umbrella organization for the ADA and NDC and was charged with increasing dairy farmer incomes by increasing dairy product sales through a coordinated promotion program. The new organization consolidates funding from all three organizations. The UDIA has become a federation of state or regional dairy promotion organizations (called "member organizations"). In 1995, for example, the UDIA was made up of 18 member organizations. The UDIA's annual budget is split between advertising and promotion, research, and education. In 1989, for example, the UDIA's budget was $40 million, of which 77.4 percent was for advertising and marketing services, 11.8 percent was for nutrition research and education, 2.6 percent was for product/process research and development, 5.3 percent was for program support, and 2.9 percent for other uses (UDIA 1989).

In 1983, Congress passed the Dairy and Tobacco Adjustment Act. This act authorized a national program for dairy product promotion, research, and nutrition education as part of a comprehensive strategy to increase human consumption of milk and dairy products and to reduce surpluses. Congress was concerned about the rising cost of dairy surpluses and wanted the dairy industry to take on more responsibility for it. It required mandatory funding in the form of a 15-cents-per-hundredweight checkoff for all milk marketed in the continental United States. The National Dairy Promotion and Research Board (NDPRB) was formed from this authorization and receives 5 cents of the 15-cent checkoff funds. The NDPRB has three main responsibilities: (1) to conduct national promotion and research programs with the check-off funds, (2) to coordinate NDPRB programs with existing programs of state and regional promotion organizations, and (3) to evaluate the check-off program and to report an evaluation to Congress.

Ten cents of the check-off funds can go to any approved local or regional promotion organization. Dairy farmers that are members of regionally affiliated UDIA member organizations can have 10 cents of their check-off dollars directed to the UDIA. Nonmember dairy farmers in states such as Wisconsin, California, Oregon, and Washington can elect to have 10 cents of their check-off dollars directed to their state or regional promotion organization.

The NDPRB and UDIA recently formed Dairy Management Incorporated (DMI) in March 1994 after recognizing duplication in promotional efforts and the need to better coordinate scarce promotion dollars. The new organization was formed as a response to pressures on the

NDPRB and UDIA to become more efficient. The purpose of DMI is to improve joint planning, to increase program and cost effectiveness, to improve industry communications, and to enhance accountability by combining the staffs and program administration of both organizations. Under this arrangement, DMI is responsible for development and execution of all national marketing programs for milk, cheese, butter, and frozen desserts paid for by check-off dollars. DMI is also responsible for other industry functions including product research and development, nutrition education, public relations, market research, and development of export markets.

Fluid milk processors have also become involved in the advertising and promotion of milk. The National Fluid Milk Processor Promotion Board (NFMPPB) was established in 1994. The 1990 Farm Bill established the authority for this program, and Congress extended authority in 1996.

In October 1993 all fluid milk processors processing over 500,000 pounds of fluid milk per month were eligible to vote in a referendum to support the establishment of a fluid milk promotion program. The referendum passed with 72 percent of all processors who represent 77 percent of the volume of fluid milk in the United States. Processors were then assessed 20 cents per hundredweight for a six-month period to fund a trial educational and marketing program.

The NFMPPB was established with the goal of changing consumers' attitudes about fluid milk and ultimately reversing the long-term per capita decline in fluid milk sales. It is the only program of its kind to promote the benefits of skim and low-fat milks. The secretary of agriculture seated the board in mid-1994. It is made up of 20 members, 15 of whom are milk processors representing different geographic regions. The other 5 members are at large, one of whom is a public member and cannot be a fluid milk processor.

The board has overseen the development and implementation of a fully integrated program that consists of advertising, public relations, and promotion to educate consumers about milk. The cornerstone of the campaign is the now-famous milk-mustache ads. Research conducted on the effectiveness of the campaign by Roper Starch Worldwide has shown that the campaign has had an impact on consumer attitudes about milk and has corrected some of the misconceptions. In March 1996 processors voted to continue the campaign, with another referendum to be requested in 1998.

NOTES

1. "Milk and Cream," *Code of Federal Regulation,* title 21, pt. 131, 1995 ed.

2. The Pasteurized Milk Ordinance requires that raw milk be cooled to at least 45°F within two hours after milking and must be maintained at the temperature until delivered.

REFERENCES

Campbell, John R. and Robert T. Marshall. 1975. *The Science of Providing Milk for Man.* New York: McGraw-Hill Publishing Co.

Giblin, James. 1986. *The Fight for Purity.* New York: Crowell.

Henderson, James Lloyd. 1971. *The Fluid-Milk Industry.* Westport: AVI Publishing Company, Inc.

International Dairy Food Association. 1996. *Hot Line,* vol. 7, No. 3 (M), November 21.

Lough, Harold W. 1991. *Fluid Milk Processing Market Structure.* Station Bulletin 619. Department of Agricultural Economics, Purdue University, August.

Manchester, Alden C. 1983. *The Public Role in the Dairy Economy: Why and How Governments Intervene in the Milk Business.* Westview Special Studies in Agricultural Science and Policy. Boulder: Westview Press.

Milk Industry Foundation. 1995. *Milk Facts: 1995 Edition.* Washington, D.C., October.

Putnam, Judith J. and Jane E. Allshouse. 1996. *Food Consumption, Prices, and Expenditures,* 1970–94. Statistical Bulletin 867. U.S. Department of Agriculture, Economic Resource Service, April.

Prepared Foods. 1995. *A New Life for Borden.* July.

Selitzer, Ralph. 1976. *The Dairy Industry in America.* New York: Dairy Field and Ice Cream. Books for Industry, Division of Magazines for Industry.

Tauer, Janelle R. and Olan D. Forker. 1987. *Dairy Promotion in the United States, 1979–86: The History and Structure of the National Milk and Dairy Product Promotion Program with Special Reference to New York.* AE Research 87-5. Department of Agricultural Economics, Cornell University, June.

United Dairy Industry Association. 1989. *UDIA 1989 Annual Report: Meeting the Challenge.*

U.S. Department of Agriculture. Economic Research Service. 1996. *Food Consumption, Prices, and Expenditures, 1970–94.* Statistical Bulletin 928. April.

U.S. Department of Health and Human Services. Public Health Service. Food and Drug Administration. 1993. *Grade "A" Pasteurized Milk Ordinance: 1993 Revision.* Public Health Service/Food and Drug Administration Publication 229.

U.S. Department of Health and Human Services. Food and Drug Administration. 1996. *Final Rule by the Food and Drug Administration, HHS.* Federal Register, November 20, 1996, vol. 61, no. 225, pp. 58991–59001.

Chapter 4

Soft Manufactured Dairy Products

Soft manufactured dairy products such as ice cream, evaporated and condensed milk, and cottage cheese accounted for almost 10 percent of all milk used in 1994 on a milkfat basis (National Dairy Promotion and Research Board 1995). By far the largest single use of milk in this category was ice cream, accounting for 8.2 percent of all milk used that year. The next highest category was canned milk (0.8 percent). Creamed cottage cheese and bulk condensed whole milk accounted for 0.3 percent each. This estimate, however, understates the total amount of milk used since it is based on a milkfat estimate. Many new low-fat products are not accounted for here such as low-fat ice cream, yogurt, cottage cheese, and frozen yogurt and condensed and evaporated milk made from skim milk. While soft manufactured dairy products have a longer shelf life than packaged fluid milk products that aren't ultrapasteurized, they are not as storable as hard manufactured dairy products such as butter, nonfat dry milk, and cheese.

Ice Cream

The first ice cream was made somewhere in Europe in the fifteenth or sixteenth century, possibly in Italy or maybe England. It was developed by chilling juices and wines and using primitive water ices. Later ice cream was made by chilling concoctions containing milk and cream (Dickson 1972). It has been recorded that Nero Caesar, the fifth Roman emperor (54–68 A.D.), sent his best runners into the mountains for snow in order to combine it with honey, juices, and fruit pulps. Frozen desserts were first brought to France by Catherine de Médicis in 1533 when she left Italy at 14 to marry the Duke of Orleans. She introduced ices and sherbets to France via her cooks and chefs.

Ice cream was later brought to the early American colonies. A letter written in 1700 by a guest of Governor Bladen of Maryland described being served ice cream (Frandsen and Arbuckle 1961). By the time of the Revolution, ice cream was served in exclusive confectionery shops in New York (Selitzer 1976). Philip Lenzi advertized the sale of ice cream in the May 12, 1777, edition of the *New York Gazette*. A confectioner from London, Lenzi was trying to appeal to the British citizens who made their headquarters in New York. George Washington was reported to have two pewter ice cream pots. And Thomas Jefferson produced an 18-step recipe for making and serving ice cream.

Dickson writes that for many years, ice cream was a dish "of the very wealthy in America, an eighteenth-century rarity and delicacy of the highest order." Ice cream was really popularized in America after a number of technological advancements that made it plentiful and affordable for all. The first technological innovation was insulated ice houses where ice was harvested from lakes and streams and stored throughout the summer and fall months. Ice cream then became available at all times of the year, not just when it snowed or hailed. Ice cream was made during this period by the pot freezer method, whereby a mixture of milk, cream, and other products were beaten in a pot that was shaken up and down in a pan of salt and ice. The salt lowered the temperature of the water-ice mixture. The next technological innovation was developed by the wife of a young naval officer, Nancy Johnson, who invented the hand-cranked ice cream freezer. Her novel idea was later patented by W.G. Young on May 30, 1848. After the turn of the century, a number of technological innovations created the modern ice cream industry we see today: first practical steam power, then mechanical refrigeration, the homogenizer, electric power and motors, new testing equipment, packaging machines, freezers and freezing processes, new insulation concepts, and the motorized delivery van (Dickson 1972). These innovations allowed ice cream production to expand economically to meet growing demand.

Ice cream production grew from 5 million gallons a year in 1899 to 150 million gallons in 1919. During this period ice cream was sold in confectionery shops or by hokeypokey men who sold cheap ice cream from a pushcart. In the 1920s, Americans love for ice cream soared as ice cream plants began to be modernized and ice cream was delivered by motorized ice cream trucks. Novelties such as Eskimo Pies and Good Humor, a candy sucker lollipop on a wooden stick, were introduced and became popular. By 1930, ice cream production reached 277 million gallons and was the nation's fastest-growing industry. But the Great Depression and the repeal of Prohibition reduced ice cream sales and production to 162 million gallons by 1933. According to Selitzer (1976), it was during these hard times that the industry saw the beginnings of retail ice cream chains.

Ice cream was also a favorite morale booster in wartime. During the Civil War, Jacob Fussell, the father of the wholesale ice cream business, sold ice cream to Union supply officers from his Washington plant. Ice cream was endorsed by the U.S. Army during World War I as a morale food and was made available to American troops. In World War II, Selitzer (1976) reports that the Navy built a floating ice cream parlor at a cost of $1 million. "A concrete barge was outfitted to produce ten gallons of ice cream a second." During the war, ice cream was more effective in boosting morale than beer. Dickson (1972) writes that "ice cream was at once an edible symbol for good morale, for what was being fought for, and for America itself."

By the end of World War II, as food rationing ended, ice cream production exploded to meet strong consumer demand. The 1950s then saw the end of the drugstore soda fountain and the beginning of retail supermarkets and ice cream store chains.

Today, ice cream sales are big business. Retail sales of ice cream and related products reached a new record of $10.5 billion in 1994 (International Ice Cream Association 1995). Of that amount, $5.2 billion was for away-from-home sales, and $5.3 billion for at-home sales. Sales of frozen desserts were $3.4 billion for ice cream, $1.4 billion for ice milk, $1.7 billion for frozen yogurt, and $4 billion for all others. Over the past 30 years, per capita consumption of ice cream and related products has stabilized at just over 23 quarts per person per year. This was a significant increase from 17 quarts per person in 1950 and just 6 quarts per person in 1919 (Selitzer 1976).

Standards and Classifications

The definition of what constitutes a frozen dessert and the ingredients used to make it are published each year in the *Code of Federal Regulations*

(*CFR*).[1] This collection of federal standards of identification was recently revised by the Food and Drug Administration due to passage of the Nutrition Labeling and Education Act (January 6, 1993). Called NLEA for short, this act amended standards of identification for various foods and established labeling regulations for packaged products. The NLEA also gave federal standards preemption over state standards. The NLEA revised the definition of what constitutes food products labeled "light," "lite," "reduced fat," "low-fat," and "fat-free".[2] The content of ice cream products was therefore changed to meet these new criteria for content. Ice cream products with the label "light" or "lite" must derive 50 percent or more of their calories from fat, and their fat content must be reduced by 50 percent or more. Ice cream products with the label "fat-free" must contain less than 0.5 grams of fat per serving and must have no added ingredients that contain fat. The terms "low-fat" or "low in fat" may be used if the product contains 3 grams or less of fat per serving. Standards are important to the dairy industry since they ensure that any product labeled "ice cream" meets certain criteria for content. These standards guarantee that consumers' expectations are met.

Ice cream is defined in the *CFR* as a food produced by freezing, while stirring, a pasteurized mix consisting of one or more specific dairy ingredients, optional caseinates, and other safe and suitable non–milk-derived ingredients. These dairy ingredients may be, for example, milk, butter, evaporated and condensed milk, dried milk, skim milk, nonfat dry milk, and sweet cream buttermilk. Ice cream may be sweetened with nutritive carbohydrate sweeteners and may or may not be characterized by the addition of flavoring ingredients. Frozen custard is similar to ice cream but contains egg yolks and additional flavorings and color. Campbell and Marshall (1975) note that major components of ice cream are milkfat, nonfat milk solids, sweeteners, stabilizers, emulsifiers, and flavoring. Most ice cream is sweetened with sucrose, beet extract, cane sugar, or hydrolyzed corn starch. Stabilizers are used to lessen the crystallization of ice and hence to increase the smoothness of the product. Emulsifiers are used to provide the uniform whipping quality of the mix (Frandsen and Arbuckle 1961). Ice cream must contain not less than 1.6 pounds of total solids to the gallon and must weigh not less than 4.5 pounds to the gallon. It must contain not less than 10 percent milkfat and not less than 10 percent nonfat milk solids. Nonfat milk solids can only be less than 10 percent when the milkfat content is greater than 10 percent.

Mellorine is an imitation frozen dairy dessert that replaces milkfat with animal or vegetable fat. Mellorine was introduced back in the 1960s as a substitute to true ice cream and never really became popular. The *CFR* requires that mellorine contain not less than 1.6 pounds of total solids to the gallon and weigh not less than 4.5 pounds to the gallon. It must contain not less than 6 percent fat.

Ice Cream Processing

The process of manufacturing ice cream has changed very little over the past 50 years. The first step is to develop an ice cream mix, or recipe, that contains all of the ingredients needed to make a particular ice cream product. After the recipe is made, all of the ingredients are combined in proper proportions in a mixing tank. This liquid mix must meet the minimum standards for the *CFR;* thus laboratory tests must be completed on the finished mix (Campbell and Marshall 1975). After the mix is made and all of the ingredients are combined, it must be pasteurized by vat, high-temperature short-time, or ultrahigh-temperature methods according to standards set by the Food and Drug Administration. The most common type of pasteurization is the high-temperature short-time method in which the mix is heated to 175°F (79°C) for 25 seconds (International Ice Cream Association 1995). After the mix is pasteurized, it is homogenized under pressures of 2,000 to 2,500 pounds per square inch in order to break up the milkfat globules and to prevent cream separation. After homogenization, the mix is quickly cooled to a temperature of about 40°F (4.4°C) in order to prevent microbial growth and to retard rates of chemical reactions. The mix usually is aged three to four hours to allow the milk proteins and added stabilizer to tie up free water, which aids mix viscosity development.

The mix is next frozen by one of two methods: a continuous freezer, which freezes a steady flow of mix into ice cream, or a batch method, which makes a single batch of ice cream at one time. In both methods, while the mix is being frozen, it is whipped and aerated by blades in a freezer called "dashers." It is the combination of the whipping and freezing that produces ice cream. The mix is aerated in order to allow the finished ice cream to have desirable eating characteristics. Without air incorporated into the mix, ice cream would be very hard and unmanageable. After the ice cream is made, it is still relatively soft and is packaged in containers during filling operations. Ingredients such as fruits and nuts may be added at this time. Packaging may include large containers; half-gallon, quart, and pint packages; paper cups; or molds for ice cream on a stick (International Ice Cream Association 1995). After it is packaged, it is sent to a hardening room where it is stored at a temperature less than –20°F (–29°C). Ice cream is hardened rapidly to prevent ice crystal formation, where ice cream melts and then rehardens during transit from the dairy processing facility to the consumer's home.

Table 4.1. Ice cream and related products: total and per capita production

	Ice cream		Ice milk[a]		Frozen yogurts[b]	
Year	Total (1,000 gal)	Per capita (qt)	Total (1,000 gal)	Per capita (qt)	Total (1,000 gal)	Per capita (qt)
1940	318,088	9.64	10,457	0.32		
1950	554,351	14.66	36,870	0.98		
1955	628,525	15.30	90,185	2.20		
1960	699,605	15.55	145,177	3.23		
1965	757,000	15.65	230,992	4.77		
1970	761,732	14.95	286,663	5.63		
1975	836,552	15.53	298,789	5.55		
1980	829,798	14.61	293,384	5.17		
1985	901,449	15.16	301,312	5.07		
1986	923,597	15.38	314,673	5.24		
1987	928,356	15.33	327,561	5.41		
1988	882,079	14.43	354,831	5.80		
1989	831,159	13.47	376,507	6.10	82,454	1.34
1990	823,601	13.21	352,271	5.65	117,577	1.89
1991	862,638	13.68	341,793	5.42	147,137	2.33
1992	866,110	13.58	328,185	5.15	134,067	2.10
1993	866,248	13.42	325,346	5.04	149,933	2.32
1994	876,434	13.47	359,895	5.53	150,795	2.32

Source: International Ice Cream Association 1995.

[a]Includes freezer-made milkshakes.

[b]Frozen yogurt production data was not collected by the USDA prior to 1989. The USDA increased the overrun factor for frozen yogurt in 1991, which may have resulted in an overstatement of production for that year.

Production and Consumption

Ice cream production expanded significantly following World War II. Total production grew from 318 million gallons in 1940 to a peak of 928 million gallons in 1987 (Table 4.1). Production thereafter declined to 824 million gallons in 1990 and grew to 876 million gallons by 1994. On a production per capita basis, which takes into consideration population growth, production peaked at 15.65 quarts per person in 1965 and in general declined thereafter. Production of ice milk has grown steadily since the mid-1960s due to health concerns. It peaked at 6.10 quarts per person in 1989. Production of frozen yogurt, a low-fat ice cream–like product made from yogurt and low-fat ice cream, has grown rapidly in the late 1980s from 1.34 quarts per person to 2.32 quarts in 1994.

Ice cream and other frozen products are produced year-round. Production is highest from March through September (Figure 4.1). The peak production months of the year are June through August. Production is therefore clearly correlated with warm weather and greater leisure time and is not necessarily related to the milk production cycle, which peaks in April and May.

Ice cream and related products were produced in over 500 plants in virtually every state in the nation in 1993. The top six leading production states for ice cream in 1994 were California, Pennsylvania, Texas,

Figure 4.1. Monthly production of ice cream and related products, 1993

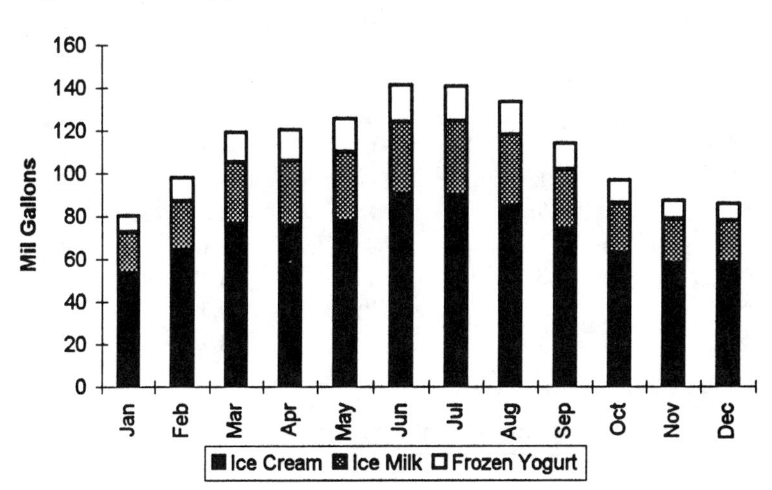

Indiana, Ohio, and New York. These states have two things in common: (1) they are major dairy states, and (2) they have or are adjacent to states with large population centers. Like fluid milk, ice cream processing plants are often located near large urban population centers.

Per capita consumption of frozen dairy products has declined over the years due to changing consumer preferences and demographics. Ice cream consumption declined from 18.5 pounds per person in 1975 to 15.8 in 1990 and has rebounded slightly to 16.1 pounds in 1994 (Table 4.2). Some of this rebound may have been due to the recent shift in consumer

Table 4.2. Per capita consumption of ice cream and related products, pounds

Year	Ice cream	Ice milk	Sherbet	Other frozen dairy products[a]
1975	18.5	7.7	1.5	0.3
1976	17.9	7.3	1.5	0.3
1977	17.5	7.7	1.5	0.3
1978	17.4	7.7	1.4	0.4
1979	17.1	7.3	1.3	0.3
1980	17.5	7.1	1.3	0.3
1981	17.4	7.0	1.3	0.6
1982	17.6	6.6	1.3	0.6
1983	18.1	6.9	1.3	0.6
1984	18.2	7.0	1.3	0.8
1985	18.1	6.9	1.3	1.5
1986	18.4	7.2	1.3	1.1
1987	18.4	7.4	1.3	1.2
1988	17.3	8.0	1.3	1.2
1989	16.1	8.4	1.3	2.9
1990	15.8	7.7	1.2	3.7
1991	16.3	7.4	1.1	4.3
1992	16.3	7.1	1.2	4.4
1993	16.1	6.9	1.3	5.0
1994	16.1	7.6	1.4	4.9

Source: Dairy Management, Inc.

[a]Includes frozen yogurt and nonfat ice cream after 1989.

preferences toward superpremium ice cream, a quality high-fat ice cream, whose sales volume increased 6 percent in 1993 (International Ice Cream Association 1995). Ice milk consumption has declined from 7.7 pounds per person in 1975 to 6.6 pounds in 1982, then rebounded to 8.4 pounds in 1989. It has since fallen to 6.9 pounds in 1993 and grew to 7.6 pounds in 1994. Some of that rebound was likely due to consumer concerns about cholesterol in the diet. That concern helped increase the demand for frozen yogurt and nonfat ice cream, whose consumption increased steadily from 2.9 pounds per person in 1989 to 5 pounds in 1993. Per capita consumption of sherbet, a nondairy frozen dessert that competes with frozen dairy desserts, declined very slightly over the period 1975–93.

U.S. export of ice cream products has increased threefold in recent years (Table 4.3), from $27.7 million in 1990 to $82.8 million in 1994 (International Ice Cream Association 1995). On a volume basis, exports rose nearly 18 percent in 1994 from the year before. The top three regions of the world that showed the most growth in the last five years are North America (up 246 percent), Europe (up 169 percent), and the Far East (up 364 percent). The top five countries in terms of sales were Japan, Mexico, Russia, France, and Hong Kong. The single largest market in 1994 was Japan, with sales of $27.7 million, a 33 percent increase over the year before. Mexico was in second place with 1994 sales of $13.9 million, which increased 38 percent from the year before. The Russian federation has emerged as a significant market for the United States. Sales grew to $6.8 million in 1994, up a whopping 324 percent from 1993 levels.

Yogurt

Due to strong consumer demand, yogurt sales in the United States grew significantly in the last 30 years from 169 million pounds in 1970 to 1,228 million pounds by 1994 (USDA, ERS 1994). The wholesale value of packaged yogurt rose dramatically from $685 million in 1987 to almost $1.2 billion by 1994 (Milk Industry Foundation 1995). Yogurt's popularity can be attributed to its image as a healthy natural product, improved marketing, the addition of fruits and sweeteners to the product, a switch to lower-fat yogurts, and the nutritional value of the product. Yogurt contains large quantities of protein, lactose, and B vitamins and is highly digestible (Campbell and Marshall 1975).

Table 4.3. U.S. exports of ice cream, tons

Region/country	1990	1991	1992	1993	1994
North America	2,418	4,414	5,635	6,886	9,870
Canada	380	600	768	712	431
Mexico	1,561	3,500	4,670	5,908	9,192
Bermuda	477	314	197	266	247
Caribbean	1,090	1,045	744	863	985
Netherlands Antilles	604	652	320	372	429
Other Caribbean	486	393	424	491	556
Central America	0	13	106	106	140
South America	112	67	332	279	350
Europe	2,764	6,364	12,503	5,828	6,180
United Kingdom	1,435	2,606	3,607	1,916	1,685
France	587	1,916	30	2,898	1,763
Germany	358	716	27	21	170
Greece	310	459	248	16	64
Netherlands	4	436	8,387	122	7
Other EC-12	0	0	34	41	15
Eastern Europe	0	14	2	48	43
Soviet Union	70	202	113	705	2,374
Other Europe	0	15	55	61	59
Middle East	1,008	1,218	1,261	711	1,906
Far East	4,085	7,009	10,940	15,369	16,398
Singapore	198	336	357	674	487
Indonesia	14	76	65	227	229
South Korean	15	66	687	415	894
Hong Kong	1,312	1,503	2,101	3,600	2,734
Taiwan	145	295	257	256	752
Japan	2,312	4,552	7,168	9,791	10,243
Other Far East	89	181	305	406	1,059
Oceania	1,205	1,582	1,056	804	472
Australia	463	1,122	739	534	242
Other Oceania	742	460	317	270	230
Sub-Saharan Africa	18	101	7	18	20
World	12,700	21,813	32,584	30,864	36,321

Source: Dairy Management, Inc.

Fermented milk has been eaten for many centuries. Rašić and Kurmann (1978) trace the origins of yogurt to Asia and the Balkans. The name *yoghurt* is believed to have been coined by nomadic Turks in the eighth century. In his travels in 1712, E. Kaempfer described a special room where yogurt was made in the palace of the emperor of Persia. The earliest inhabitants of the Balkans used to breed large flocks of sheep and made soured milks called *Prokish,* which later became known as yogurt. The Slavs, who later occupied the region, adopted the preparation of yogurt. In 1995 this author visited a small sheep farm in the mountains of Macedonia in the southern Balkans and was served a light, refreshing liquid yogurt that was prepared from sheep's milk. This recipe no doubt had its origins from Prokish.

In the late nineteenth century, modern-day yogurt gained popularity in Europe due to Nobel prize winner Professor Éli Metchnikoff (Selitzer 1976). Metchnikoff, who was director of the Pasteur Institute, toured the Rodope Mountains of Bulgaria and noted longevity was common among the mountain people. He surmised it was attributable to the native food yogurt. He developed his theory of longevity and linked it to the beneficial effect of yogurt in the diet. Yogurt was then promoted as a health food in Europe and was often prescribed by doctors to their patients.

Professor Metchnikoff's work was noticed by the Spanish businessman Isaac Carasso. Carasso acquired yogurt cultures from Bulgaria and began to produce yogurt in Barcelona, Spain. His business grew, and he decided to extend his activities to France. It was there that he placed his son Daniel in charge of the operation. Carasso named his new product Danone after his son. When Hitler's armies began their persecutions, Daniel moved to the United States and started a new company, the Dannon Company, in New York City in 1942. Dannon today is one of the largest manufacturers of yogurt in the United States.

Standards and Classifications

Yogurt is defined in the *CFR* as the food produced by culturing cream, milk, partially skimmed milk, or skim milk (alone or in combination) with a characterizing bacterial culture that contains the lactic acid–producing bacteria *Lactobacillus bulgaricus* and *Streptococcus thermophilus.*[3] Other dairy products, such as nonfat dry milk, may be used to increase the solids-not-fat content of the product. The *CFR* classifies three types of yogurts that can be marketed in the United States: yogurt, low-fat yogurt, and nonfat yogurt. Yogurt can be marketed as is or frozen and whipped into a dairy dessert.

Yogurt must contain not less than 3.25 percent milkfat and not less than 8.25 percent solids-not-fat before the addition of bulky flavors. In addition, it must have a titratable acidity of not less than 0.9 percent, expressed as lactic acid. It may be homogenized but must be pasteurized or ultrapasteurized before the addition of the bacterial culture. The definition of low-fat yogurt is similar to that of regular yogurt except that it must contain between 0.5 and 2 percent milkfat and not less than 8.25 percent solids-not-fat before the addition of bulky flavors. Nonfat yogurt contains less than 0.5 percent milkfat and not less than 8.25 percent solids-not-fat before the addition of bulky flavors. While the *CFR* requires a minimum solids-not-fat content of 8.25 percent, most yogurt marketed in the United States contains 12 to 15 percent total solids (both milkfat and solids-not-fat) since a higher solids content increases the firmness of the product (Campbell and Marshall 1975).

Yogurt Processing

The basic ingredients of yogurt are raw milk and bacteria cultures (Rašić and Kurmann 1978). The milk is converted to yogurt by growing within it certain strains of lactic acid bacteria and allowing the milk to sour under carefully controlled conditions. The first step in processing yogurt is to prepare the raw milk. It must be clarified, standardized, homogenized, and pasteurized. Other dairy ingredients can be added to milk after standardization in order to increase the solids-not-fat content of the milk. Sugars, sweeteners, and stabilizers may also be added. Stabilizers are used to stabilize the lactic acid gel and improve the consistency of the product. Heat treatment of the milk to 185° to 190°F (85° to 88°C) for 30 to 45 minutes is critical since it destroys all random bacteria that may be present in raw milk, inactivates naturally occurring enzymes, and improves the consistency of the yogurt (Hargrove 1970). Heat treatment also denatures the whey proteins present in milk, which promotes the growth of the bacteria.

After the milk is properly processed, it is moved into a large fermentation vat to be further processed into yogurt. There the pasteurized milk is cooled to a temperature of 113°F (45°C) and is inoculated with two strains of bacteria in fixed proportions: *Lactobacillus bulgaricus* and *Streptococcus thermophilus*. The mixture is then packaged within 30 minutes of inoculation into cups and tubs. After packaging, it is further cooled to 106° to 108°F (41° to 42°C) and allowed to incubate for a period of three to eight hours until the desired pH and characteristic consistency of yogurt is reached. After the desired acidity has been reached, the yogurt is wheeled into a cold room where the incubation process is stopped (Henderson 1971).

A vat method is also used for viscous or semiliquid yogurt. Here the inoculated milk is transferred to a large vat where the yogurt is allowed to incubate and set up. It is then cooled, stirred, and transferred to the filler machine, where it is packaged.

After the yogurt is made, it is mixed with sugar, fruit flavoring (if required), and in some cases preservatives. Two styles of fruit-flavored yogurt are common in the United States: Swiss-style, which mixes the fruit into the yogurt, and sundae-style, which places fruit on the bottom of the cup. Flavors are added to the yogurt in one of three ways depending on style (Hargrove 1970): (1) mix blending, whereby the fruit and flavoring are added to the mix prior to fermentation, (2) fruit on the bottom, where preserves or fruit are added to the package before filling and setting, and (3) bulk mixing, where fruit and flavorings are blended into the yogurt after fermentation.

Production and Consumption

Yogurt consumption does not account for a large proportion of milk produced in the United States. In fact, yogurt accounted for just 1.8 percent of all fluid milk and cream sales in federal orders in 1993 (Milk Industry Foundation 1995). Nevertheless, production of yogurt has grown in recent years from 982.6 million pounds in 1990 to 1,379 million pounds in 1994 (Dairy Management, Inc. 1995). Consumption trends for yogurt show a significant rise starting in the mid-1980s. Per capita consumption doubled over the past 20 years, increasing from 2.1 pounds in 1975 to 4.4 pounds in 1993. As a result, yogurt sales grew from a total of 324 million pounds in 1974 to 1,228 million pounds in 1994.

Selitzer (1976) cites yogurt as a new product whose healthful and natural image fits today's emphasis on nutrition. The Dannon Company promoted this image by adding real fruits and preserves to yogurt in the United States starting in the 1970s. (In France Carasso first introduced the idea of adding strawberries to yogurt as early as the 1930s.) In addition, frozen yogurt has also became popular, particularly when dispensed as a soft-serve alternative to high-fat ice cream. Significant quantities of yogurt have also been exported in recent years. Yogurt exports rose from $6.9 million in 1990 and peaked at $14.4 million in 1992, then fell to $11.4 million in 1994. The top five U.S. export customers in 1994 were Mexico (5,722 metric tons), Japan (454 metric tons), Canada (374 metric tons), Norway (220 metric tons), and Australia (186 metric tons).

Evaporated and Condensed Milk

A major problem with milk prior to pasteurization and refrigeration was that it had to be consumed directly after it was purchased because of its perishability. In addition milk at the time was often inferior since it either was hauled long distances from the country to the city via train, was unrefrigerated, or was produced by the swill dairies. Thus consumers in the 1800s could not purchase quality milk for consumption at a later time.

Gail Borden became interested in the idea of preserving food without refrigeration so that it could be consumed when needed. His first attempt was a dehydrated meat and flour biscuit. Borden envisioned wide use of the product by armies, navies, and explorers. Unfortunately, the product tasted putrid and was not a commercial success. Then Borden began to experiment with condensing and preserving milk. Selitzer (1976) writes that Borden came up with the idea after a particularly rough ocean voyage during which two dairy cows used to feed infants became seasick and could not produce milk. As a result, several infants who depended on the milk died.

Borden was not the first person to successfully can and preserve milk. In the early 1800s, Nicholas Appert was; he successfully sterilized milk in bottles. Borden, however, was the first to retain the color and taste of preserved milk. He learned of an unusual method of preserving food in a Shaker community in New Lebanon, New York. He took the Shakers' equipment and applied it to condensing and preserving milk. The milk was heated and slowly reduced to three-fourths of its original volume by a partial vacuum that removed the moisture at relatively low temperatures. This method preserved color and produced a product superior in flavor to milk boiled in an open vat (Hall and Hedrick 1966). Borden placed his product on the market and called it "plain and condensed milk." While it was hermetically sealed in cans, it would only stay fresh for a week. He later added sugar as a preservative and marketed it as sweetened and condensed milk starting in 1861.

Borden's new product didn't become a commercial success until the ill effects of swill milk were well known. Mothers turned to Borden's product since he used fresh country milk produced under strict, sanitary conditions. Then, with the advent of the Civil War, the Union army purchased Borden's milk to feed its troops. It was popular since it could be carried for long periods of time and would remain fresh. Borden was succeeded by his son John, who helped build the dairy company.

Johann Meyenberg began to experiment with new methods to preserve milk without the use of sugar in the late 1800s in Switzerland. Meyenberg worked for the Page brothers, who took Borden's idea for sweetened and condensed milk to Europe. Meyenberg then resigned from his position in Switzerland, moved to St. Louis, and began to experiment with a new method for preserving milk. This method sterilized milk under pressure while the cans were agitated, and it did not require sugar to preserve the product. It was first called "evaporated cream" but is known today as evaporated milk. Evaporated milk is still produced in much the same way developed by Meyenberg. After perfecting his method and applying for a patent, Meyenberg formed the Helvetia Milk Condensing Company in 1885, which later became known as the Pet Milk Company. Later Meyenberg went into business with E.A. Stuart and Tom Yerxa, and in 1899 they started production of evaporated milk for their new business, the Carnation Company. Both Pet Milk and Carnation have become giants in the evaporated milk business.

Standards and Classifications

The *CFR* describes sweetened condensed milk as the food obtained by partial removal of water from a mixture of milk and safe and suitable nutritive carbohydrate sweeteners. The finished product should contain not less than 8 percent by weight of milkfat and not less than 28 percent by weight of total milk solids. The product must be pasteurized and may be homogenized. Sweetened condensed skimmed milk is made from skim milk and contains not more than 0.5 percent milkfat by weight (unless otherwise indicated) and not less than 24 percent by weight of total milk solids.

Evaporated milk is described in the *CFR* as the liquid food obtained by partial removal of water from milk. It should contain not less than 6.5 percent milkfat by weight, not less than 16.5 percent of solids-not-fat by weight, and not less than 23 percent of total milk solids by weight. Evaporated milk contains added vitamin D (addition of vitamin A is optional), is homogenized, and is sealed in a container. It is processed by heat, either before or after sealing, so as to prevent spoilage. Evaporated skimmed milk is made from skim milk and contains not less than 20 percent total milk solids by weight and not more than 0.5 percent milkfat by weight (unless otherwise noted). Evaporated skimmed milk may be homogenized and contains added vitamins A and D.

Evaporated and Condensed Milk Processing

Manufacturing sweetened condensed milk and evaporated milk involves two separate processes. For condensed milk, water is removed from the milk under a vacuum. The vacuum allows for milk to be heated and condensed at a lower temperature than in the open vat method in order to improve taste and quality. Sugar is used to preserve the product. Evaporated milk, on the other hand, receives high preheat treatment and is actually sterilized in the can and therefore does not need sugar as a preservative.

Production and Consumption

Both condensed and evaporated milk are processed into two types of products: bulk and canned milk. Today, canned products are intended mainly for home use as a baking ingredient. But large amounts of milk are condensed in bulk quantities in order to lower the cost of transportation. Bulk milk is used or sold for further processing into other dairy products or for processing into nondairy products. For example, a butter plant or an ice cream plant may elect to purchase bulk tanker loads of condensed milk for processing from a surplus region whenever local production is short. Production of condensed and evaporated milk in recent years is presented in Table 4.4. The data indicate that almost

Table 4.4. Production of canned and bulk milk, 1990–94, pounds

	1990	1991	1992	1993	1994
Canned milk	615,183	560,046	598,571	556,742	564,545
Evaporated and condensed whole milk	602,647	543,094	582,115	534,507	537,881
Evaporated skim milk	12,536	16,952	16,456	22,235	26,664
Bulk condensed milk	1,425,810	1,529,837	1,615,106	1,605,492	1,507,286
Whole milk	249,788	282,616	292,616	238,245	206,574
Sweetened	105,371	132,579	129,704	139,211	118,800
Unsweetened	144,417	150,037	162,912	99,034	87,774
Skim	1,176,022	1,247,221	1,322,490	1,367,247	1,300,712
Sweetened	44,510	125,161	126,380	97,690	56,582
Unsweetened	1,131,512	1,122,060	1,196,110	1,269,557	1,244,130
Condensed or evaporated buttermilk	37,247	39,572	46,026	46,525	34,133

Source: Dairy Management, Inc. 1995.

three-fourths of all evaporated and condensed milk was processed as bulk, with the remaining 27 percent canned.

Demand for evaporated milk really picked up during the Spanish-American War of 1898, and later during and after both World War I and World War II. Stalin, for example, received 3 million pounds of canned milk for his Red Army from the United States during the winter of 1942–43 (Selitzer 1976). Canned milk helped alleviate starvation in Europe during and after both wars. The U.S. government began to purchase surplus quantities of evaporated milk as part of its dairy price support program. The government purchased 47 million pounds of evaporated milk in 1935 and 66 million in 1940 (Rojko 1957). The government purchased 119.5 million pounds of evaporated milk in 1941 under authority of Section 32 of the Agricultural Adjustment Act of 1935. In addition to its use by troops, evaporated milk was also promoted for feeding babies. As early as the 1920s, doctors began to prescribe evaporated milk as an infant formula. Selitzer (1976) reports that studies by Dr. Marriot, of St. Louis, and Dr. Brennemann, of Chicago, found that the important nutrients in milk were maintained during processing and that the protein in milk was made more digestible to babies. Also, it was recognized that the sterilized product was far safer than liquid milk at the time.

Consumption of both condensed and evaporated milk declined over the years when infant formulas were introduced and gained wide acceptance. In addition, fresh milk became more readily available, and strict sanitary regulations improved consumers' confidence in the nation's milk supply. As a result, consumption of canned milk waned. In more recent years, per capita consumption of evaporated and condensed milk declined from 8.9 pounds in 1975 to a low of 7 pounds in 1982, but then grew to 8.2 pounds in 1993. Much of the recent growth, however, was in bulk and canned skim milk, which grew from 3 pounds per person in 1982 to 5.2 pounds in 1993. Most canned milk today is used primarily for cooking and baking.

Cottage Cheese

Cottage cheese has its origins in central Europe, where it was made fresh in farmhouses. In colonial America, cottage cheese was a popular by-product of cheese making and was made in cottages, hence the name (Kosikowski 1966). Curds were made from skimmed milk and were usually eaten that same day. In 1918 cottage cheese was promoted by the USDA as a meat substitute during World War I (Henderson 1971). At that time, skim milk was a by-product that was either fed to farm animals or thrown away. Meat, on the other hand, was in short supply. Selitzer (1976) reports that the USDA advertised at the time that 1 pound of cottage cheese had the same protein as 1.27 pounds of sirloin steak. As a result, production jumped from 28 million pounds in 1918 to almost 68 million pounds by 1926.

Cottage cheese today represents a $688 million industry, a relatively small part of the overall dairy industry. It is often produced by bottlers since equipment needs are simple, it makes good use of skim milk left after cream is separated, and it must be produced and delivered fresh to retail stores.

Standards and Classifications

The *CFR* defines cottage cheese as a soft uncured cheese prepared by mixing cottage cheese dry curd with a creaming mixture. The finished product must have at least 4 percent milkfat by weight and contain not more than 80 percent moisture. Dry curd cottage cheese does not contain the creaming mixture and contains less than 0.5 percent milkfat and not more than 80 percent moisture. Low-fat cottage cheese is prepared in the same way as higher-fat cottage cheese, except that it must have between 0.5 percent and 2 percent milkfat by weight and must have not more than 82.5 percent moisture.

Cottage Cheese Processing

Cottage cheese production is a two-step process—making the cheese curds and then creaming the curds. Unlike ripened cheeses, in which rennin enzymes coagulate milk and form curds, cottage cheese curds are created by bacteria that form lactic acid. Milk is first separated, and the skim milk is immediately pasteurized. After pasteurization, the milk is cooled to 90°F (32°C) and pumped into a cheese vat where it is mixed with a lactic acid starter, which contains the bacteria *Streptococcus lactis* or *S. cremoris*, each dominant acid producers, and *Leuconostoc citrovorum* as a dominant flavor producer (Kosikowski 1966). After about 30 to 60 minutes, rennet is added to change the texture of the curd, to increase the ability of each curd to remain intact, to enhance whey expulsion during heating, and to produce a sweeter curd. The mix is then left to incubate, or set, for about 4–16 hours depending on the type of cheese to be made. The set is complete when the milk coagulates, the acidity of the whey reaches 0.48 to 0.52 percent, and the pH reaches 4.6 (Henderson 1971).

After the cheese is set, often in one large block in the vat, the curd is cut with special wire cheese knives. These break up the coagulin to the desired curd size. The curds are then cooked in order to expedite whey removal and to improve the cheese's texture. The curds are cooked either by directly adding steam or hot water or by indirectly heading the jacket of the vat (Campbell and Marshall 1975). After about an hour of cooking, the curds are washed and cooled with water and then drained. Whey present in the cheese curds is removed during draining. The curds are then salted, and pasteurized cream is added for flavor and to improve the solids content of the product. The product is then packaged and moved into a cold room for immediate delivery to retail outlets. It normally takes about 100 pounds of fluid skim milk to make 15 pounds of uncreamed cottage cheese (Kosikowski 1966).

Production and Consumption

Until the 1970s production of cottage cheese grew steadily since its rise in popularity in the early 1920s. Production of cottage cheese peaked at 1,115 million pounds in 1972. Since then it has steadily declined to 731 million pounds in 1994 (USDA, ERS 1994). Demand also declined from 4.7 pounds per person in 1975 to 2.9 pounds in 1993. This decline is due to consumers' unfavorable image of it and to possible substitution with yogurt.

NOTES

1. "Frozen Desserts," *Code of Federal Regulation,* title 21, pt. 135, 1995 ed.
2. "Nutrient Content Claims for Fat, Fatty Acid, and Cholesterol Content of Foods," *Code of Federal Regulation,* title 21, pts. 101.56 and 101.62, 1995 ed.
3. "Milk and Cream," *Code of Federal Regulation,* title 21, pt. 131, 1995 ed.

REFERENCES

Campbell, John R. and Robert T. Marshall. 1975. *The Science of Providing Milk for Man.* New York: McGraw-Hill Publishing Co.

Dairy Management, Inc. 1995. Dairy Database.

Dickson, Paul. 1972. *The Great American Ice Cream Book.* New York: Atheneum.

Frandsen, J.H. and W.S. Arbuckle. 1961. *Ice Cream and Related Products.* Westport: AVI Publishing Company, Inc.

Hall, Carl W. and T.I. Hedrick. 1966. *Drying Milk and Milk Products.* Westport: AVI Publishing Company, Inc.

Hargrove, R.E. 1970. "Fermentation Products from Skim Milk." In *By-products from Milk.* Westport: AVI Publishing Company, Inc.

Henderson, James Lloyd. 1971. *The Fluid-Milk Industry.* Westport: AVI Publishing Company, Inc.

International Ice Cream Association. 1995. *The Latest Scoop World Wide: Facts and Figures on Ice Cream and Related Products.* 1995 ed. Washington, D.C., December.

Kosikowski, Frank. 1966. *Cheese and Fermented Milk Foods.* Ithaca, MI: Edwards Brothers, Inc.

Milk Industry Foundation. 1995. *Milk Facts: 1995 Edition.* Washington, D.C., December.

Rašić, Jeremija and Joseph Kurmann. 1978. *Yoghurt: Scientific Grounds, Technology, Manufacture and Preparations.* Copenhagen: Technical Dairy Publishing House.

Rojko, Anthony S. 1957. *The Demand and Price Structure for Dairy Products.* Technical Bulletin 1168. Agricultural Marketing Service, U.S. Department of Agriculture, Washington, D.C., May.

Selitzer, Ralph. 1976. *The Dairy Industry in America.* New York: Dairy Field and Ice Cream. Books for Industry, Division of Magazines for Industry.

U.S. Department of Agriculture, Economic Research Service. 1994. *Dairy Yearbook: Supplement to Livestock, Dairy, and Poultry Situation and Outlook.* Statistical Bulletin 889. Washington, D.C.

Chapter 5

Hard Manufactured Dairy Products

Milk in excess of fluid needs is processed into various types of manufactured dairy products. Hard manufactured dairy products such as cheese, butter, and dried dairy products play a vital role in the dairy industry by meeting the nutritional needs of consumers and providing an alternative market for surplus milk. When production is strong, milk can be processed into storable dairy products and marketed at a later time. The data in Table 2.2. indicate that 60 percent of all milk produced in 1993 was used for manufacturing purposes. Of that amount, about 83 percent was used in the manufacture of cheese and butter.

The federal milk marketing order system was designed to work in conjunction with the market for hard manufactured dairy products. It uses discriminatory pricing to set a high price for milk used for fluid purposes. In order for this system to be effective, there must be a lower-priced market that surplus milk can be diverted to. The market for hard manufactured dairy products meets this need. The price of manufacturing milk is therefore low enough to allow all Grade A milk in excess of fluid needs to be processed.

Cheese

No one really knows when cheese was first made, but it no doubt played an important role in the survival of early man. Cheese is a vital source of protein that can be safely stored and retrieved when needed. Early hunting and gathering clans probably appreciated the fact that protein could be consumed without the need to butcher their scarce supply of domesticated animals.

It is known that cheese was made and consumed in biblical times (National Cheese Institute 1995). The tomb of the ancient pharaoh Horus contained evidence of what was believed to have been cheese. In 1924 after an archeological survey, Sir Leonard Woolley concluded that inhabitants of the Fertile Crescent, situated in Iraq between the Nile on the west and Tigris and Euphrates on the east, made cheese there from the milk of cows and goats between 6000 and 7000 B.C. (Scott 1986). The Sumerians in 3500 to 3000 B.C. left behind a frieze in El-Ubaid of priests milking cows and curdling the milk into cheese (Battistotti et al. 1984; Scott 1986). In 1184 B.C., Homer referred to cheese made in caves by the Cyclops Polyphemus from the milk of sheep and goats (Scott 1986). Cheese also played an important role in the diet of ancient monarchs. When Alexander the Great defeated Darius in Damascus in 331 B.C., he found the Persian king left behind hundreds of cooks well versed in cheese making. Cheese was traded in ancient Greek markets and was part of the diet of athletes. Cheese and grain became important staples throughout the Roman Empire, which spread throughout Europe. When the Romans invaded Britain, they brought with them the art of cheese making. Palladius, a Roman in the third century, reported on cheese in Chester, which was a Roman stronghold in Britain at the time (Scott 1986).

The art of cheese making in Europe was almost lost when the Romans were conquered by the "barbarians," who did not appreciate cheese. It was the monks who kept the art alive during the Dark Ages. Cheese making then proliferated with the Vikings, who were masters of animal husbandry, and the English. It was the English who first produced cheddar cheese from skimmed milk from cows. After 1200, the Po Valley in Italy became a central market for cheese making in Europe.

At some point it was discovered that milk from the butchered stomach of a suckling animal, or inside a container made from the stomach of an animal, curdled and formed a nutritious food and pleasant-tasting drink. Animal skin bags were a common way to store liquids back in the days of nomadic tribes. Rennet, which is an enzyme made from the fourth stomach of a young cud-chewing animal, was discovered to aid the process of curdling. It reacts with casein, a major component of milk protein, causing the milk to curdle and become firm. After a certain level of acidity was reached by the action of lactic acid–producing bacteria, the lactic acid would help the rennet form curds that would then separate from the watery whey. Once pressed and stored properly, cheese was found to keep for long periods of time. Also, since cheese did contain sufficient levels of acid, few dangerous germs could live in this environment. Hence aged cheese was safe to consume and generally improved in flavor over time.

Cheese making was introduced to the New World by Sir Thomas Dale, who sailed into the Jamestown harbor with three ships and a herd of 100 cows. Cattle were scarce in the early colonies and were used primarily for milking and cheese making. Herds increased as each succeeding ship brought more cattle to the New World. By the middle of the seventeenth century dairying was taking hold along the East Coast. Selitzer (1976) reports that Delaware imported cattle from Sweden, Pennsylvania had mixed herds from Sweden and the Netherlands, and cattle in other areas were imported from Spain, England, and Denmark. New York later became the epicenter for cheese production during the mid-1800s. Cheese produced in the Goshen area of New York was shipped to many parts of the United States and exported overseas. Cheese making then spread to the Upper Midwest by the latter half of the 1800s. Many early settlers along the East Coast with roots from dairy regions in Europe, and specialized skills in the art of cheese making, moved to the Upper Midwest. There they found a climate and growing conditions similar to those of the Old World. Shortly after the turn of this century, the Upper Midwest surpassed cheese and butter production in New York and New England. These northeastern states instead focused on meeting the demands for fresh milk for a growing urban population.

The cheese industry today represents big business for dairy farmers and cheese processors. The National Cheese Institute (1995) estimates that the wholesale value of cheese products shipped in 1994 was $16.4 billion (natural cheese, $10.4 billion; processed cheese and related products, $5.3 billion; and other cheeses and cheese substitutes, $0.7 billion). The total retail value of cheese sales in 1994 was $20.9 billion. Per capita consumption of cheese increased from 17.5 pounds in 1980 to 27 pounds by 1994, adding to stability in the overall demand for milk. Much of this growth in cheese consumption was due to increased demand for Italian cheeses, particularly mozzarella for pizzas.

Standards and Classifications

Natural and processed cheeses represent two broad classifications for cheese in the United States. Natural cheeses are those cheeses that receive no further processing after being produced. Natural cheese as a category consists mainly of American-type cheeses and other natural cheeses. American cheeses are mostly cheddar, Colby, washed or stirred curd, and Monterey Jack. Other natural cheeses include Italian, blue, brick, cream, Gorgonzola, Limburger, Muenster, Neufchâtel, Swiss, Edam, and Gouda. Italian cheeses are a large part of the "other natural cheese" category and are mainly mozzarella, ricotta, provolone, Romano, and Parmesan. Processed cheeses are made by grinding, heating, and mixing natural cheeses. Typically a blend of one or more American-type cheeses are used to produce a product that has special melting properties and improved shelf life. Blended cheeses may be mixed with fruits, vegetables, and meats.

There are many unique aspects that are taken into consideration when classifying cheese. Some of these include type of cheese (such as Swiss and blue), percentage of moisture, milkfat content, age of cheese (ripe or unripe), texture, color, smell, and taste. There are many cheeses available on the market that have unique characteristics due to the way they are processed, which includes the type of microbial culture(s) used.

The most common variety of cheese made and consumed in the United States today is cheddar, which is made from an old English recipe. Cheddar cheese was named after the village of Cheddar in Somersetshire, England (Selitzer 1976). Colby cheese is another popular cheese and was first made in 1882 in Colby Township, Wisconsin (Eekhof-Stork 1976). Colby has a higher moisture content than cheddar and is milder in flavor. California Monterey Jack cheese was copied from an old monk's recipe and marketed by David Jacks starting in 1916. It is similar to Colby but is creamy white and softer.

The requirements for cheese standards in the United States are published in the *Code of Federal Regulation* (*CFR*) under authority of the Federal Food, Drug, and Cosmetic Act (21 U.S.C. 321, 341, 343, 348, 371, 379e).[1] The *CFR* does not provide a concise definition of cheese. However one definition that seems appropriate is "the product of whole, partially skimmed or fully skimmed milk, or of milk enriched by the addition of cream, which has been coagulated by lactic acid or by the addition of rennet, leading to the release and draining off of whey"

(Battistotti et al. 1984). The *CFR* provides a detailed description of 95 different cheese types. That is because each of the 95 cheeses recognized by the U.S. government is made in a slightly different manner and has a different smell, taste, and odor. For example, the *CFR* defines cheddar as having a minimum milkfat content of 50 percent by weight of solids with a maximum moisture content of 39 percent by weight. The *CFR* goes on to describe how cheddar is prepared:

> one or more ... dairy ingredients (milk, nonfat milk, or cream) may be warmed, treated with hydrogen peroxide/catalase, and is subjected to the action of a lactic acid–producing bacterial culture. One or more ... clotting enzymes (rennet and/or other clotting enzymes of animal, plant, or microbial origin) is added to set the dairy ingredients to a semisolid mass. The mass is so cut, stirred, and heated with continued stirring, as to promote and regulate the separation of whey and curd. The whey is drained off, and the curd is matted into a cohesive mass. The mass is cut into slabs, which are so piled and handled as to promote the drainage of whey and the development of acidity. The slabs are then cut into pieces, which may be rinsed by sprinkling or pouring water over them, with free and continuous drainage; but the duration of such rinsing is so limited that only the whey on the surface of such pieces is removed. The curd is salted, stirred, further drained, and pressed into forms. One or more ... other optional ingredients ... may be added during the procedure.

The USDA has a strict grading program for cheese at the processor level. Processors must meet minimum standards of inspection in order to carry the USDA grade shield. Specific standards are set for American, Monterey Jack, Colby, and cheddar. The grades for bulk American cheeses are Extra Grade, Standard Grade, and Commercial Grade. These grades are based on flavor, body, texture, color, finish, and appearance.[2] The specific grades for Monterey Jack, Colby, and cheddar are Grade AA, Grade A, and Grade B.

Cheese Processing

The steps required to process cheese are outlined by Campbell and Marshall (1975). The process begins with high-quality raw milk. It is important to have milk with a high level of casein and other milk solids in order to produce a quality product with a good yield from the milk. Milk is first standardized. This allows the processor to control the ratio of fat to protein for a particular type of cheese. After standardization the milk is clarified and pasteurized. Pasteurization kills off all microorganisms that may adversely affect the quality of the cheese. Hydrogen peroxide is sometimes used to kill off microorganisms at a lower temperature during pasteurization. This is desirable since it allows natural milk enzymes to remain active, which speeds up the ripening process. For some varieties of cheese the milk may be homogenized to add smoothness to the cheese. The milk is then immediately cooled and pumped into a cheese vat.

The next step is to set the milk, or to form the coagulum. The milk is mixed with a cheese starter and rennet. The cheese starter is a laboratory-grown lactic bacterial culture developed specifically to make cheese. It is designed to replace the microflora lost during pasteurization with a very specific culture. The batch is then heated to a temperature of 68°–104°F (20°–40°C), depending on the type of cheese to be made. The lactic culture introduced into the milk produces a limited amount of lactic acid, which in turn stimulates the clotting of the milk, or curd formation. After the curd is formed, which takes about 15 to 60 minutes,[3] it is cut with cheese knives and then cooked in order to rapidly expel the whey. The amount of time needed and the temperature used to cook the curd depends on the type of cheese being made. The shorter the cooking time and the lower the temperature, the more moist the curd and the softer the cheese.

After the curds are cooked, the whey is drained off by use of drain tables, perforated molds, or cheesecloth. The curds are then packed or knitted together—depending on the type of cheese—then salted, pressed, and ripened. Ripening is important in that it develops flavor, makes the cheese more pliable and soft, and helps rid the curd of undesirable bacteria. It actually results in the breakdown of casein into other, simpler proteins that confer a specific aroma and taste to the cheese and develop the desirable body and texture (Battistotti et al. 1984).

It is important to note that the quality of raw milk used in the manufacture of cheese can have a profound impact on cheese yield and quality. In general, the higher the milkfat and protein content of milk and the lower the somatic cell and bacteria count, the greater the cheese yield. Both milkfat and protein are vital components of cheese production. Milk low in casein—a nonsoluble protein that accounts for 78 percent of all proteins in milk—will result in low cheese yields since casein forms the structure inside which all other milk solids are trapped when the curd is being formed. Campbell and Marshall (1975) report that a high bacteria count can cause off-flavors and defective body and texture in cheese. Also, a high somatic cell count can cause inferior curd (coagulum) and, in Swiss cheese, abnormal gas holes.

Production and Consumption

The most common types of cheeses produced in the United States are cheddar and mozzarella. In 1994, cheddar production was 2.35 billion pounds, and mozzarella production 2.07 billion (Table 5.1; National

Table 5.1. Production of natural cheese, 1980–93

Year	Cheddar	All other American[a]	Total American	Total Italian	All other natural cheese[b]	Total natural cheese
			(thousand lb)			
1980	1,750,697	625,059	2,375,756	982,731	625,779	3,984,266
1981	1,933,126	709,137	2,642,263	994,398	640,900	4,277,561
1982	2,157,476	594,822	2,752,298	1,087,781	701,590	4,541,669
1983	2,351,360	576,356	2,927,716	1,200,204	691,751	4,819,671
1984	2,112,947	535,512	2,648,459	1,318,816	706,718	4,673,993
1985	2,291,656	563,574	2,855,230	1,491,314	734,398	5,080,942
1986	2,241,624	556,536	2,798,160	1,632,921	778,171	5,209,252
1987	2,284,836	431,823	2,716,659	1,799,770	827,935	5,344,364
1988	2,279,164	477,413	2,756,577	1,937,118	878,278	5,571,973
1989	2,221,204	452,871	2,674,075	2,042,894	898,403	5,615,372
1990	2,379,794	514,427	2,894,221	2,207,021	958,194	6,059,436
1991	2,266,407	502,518	2,768,925	2,328,624	957,306	6,054,855
1992	2,400,700	535,861	2,936,561	2,508,577	1,043,153	6,488,291
1993	2,376,108	581,152	2,957,260	2,494,521	1,076,391	6,528,172
1994	2,345,847	628,573	2,974,420	2,625,702	1,134,569	6,734,691

Source: National Cheese Institute 1995.

[a]Includes Colby, washed and stirred curd, and Monterey Jack.

[b]Includes blue, brick, cream, Gorgonzola, Limburger, Muenster, Neufchâtel, Swiss, part skim, full skim, and all others.

Cheese Institute 1995). Cheddar cheese accounted for 34.8 percent of all natural cheese production and 78.8 percent of total American cheese produced that year. Mozzarella cheese, on the other hand, accounted for 30.7 percent of all natural cheese production and 78.9 percent of total Italian cheese produced in 1994. Italian cheeses consist of soft varieties such as mozzarella and ricotta and hard varieties such as provolone, Romano, and Parmesan. The Italian cheeses have shown an average growth in per capita consumption of 18 percent per year from 1980 to 1993. Much of this growth has been due to mozzarella and the growth in popularity of pizza.

Cheese was produced in 449 plants across the United States in 1994 (USDA, NASS 1995). Cheese production, excluding cottage cheese, totaled 6.7 billion pounds in 1994, up 3.1 percent from the year before. The top five cheese-producing states that year were Wisconsin (2.018 billion pounds from 153 plants), California (926 million pounds from 45 plants), Minnesota (658 million pounds from 17 plants), New York (560 million pounds from 34 plants), and Pennsylvania (348 million pounds from 21 plants). In this survey, Minnesota produced the most cheese per plant (an average of 38 million pounds per plant), and Wisconsin, with the largest number of plants, produced the least amount of cheese per plant (an average of 13.8 million pounds). Cheese plants are typically located in regions of the country with milk production in excess of fluid needs. These states represent traditional dairy states. New dairy states such as New Mexico, Texas, Arizona, and Idaho are in the process of expanding their cheese-processing capacity in order to reduce transportation costs and improve prices for area dairy farmers.

Per capita consumption of natural cheese has expanded significantly over the years, from 17.5 pounds in 1980 to 26.8 pounds in 1994. This rising growth in cheese consumption has fueled overall demand for milk and helped stabilize milk prices. In addition, it has resulted in fewer surplus dairy products ending up in government inventory. The per capita consumption of various cheese products can be seen in Table 5.2. The leading cheeses in terms of per capita consumption in 1994 were cheddar (9.11 pounds), mozzarella (7.93 pounds), processed cheese (5.29 pounds), cheese foods and spreads (3.48 pounds), and other American cheeses (2.45 pounds).

In 1994 commercial exports of cheese rose in value to $72 million, up 45 percent from the year before. While small in comparison with imports

Table 5.2. Per capita consumption of natural and processed cheese

	1990	1991	1992	1993	1994
			(lb)		
American-type cheeses	11.13	11.07	11.33	11.41	11.55
Cheddar	9.04	9.05	9.20	9.13	9.11
Other American[a]	2.09	2.02	2.13	2.28	2.44
Italian cheeses	8.97	9.37	9.96	9.82	10.28
Mozzarella	6.93	7.22	7.71	7.55	7.93
Ricotta	0.78	0.84	0.88	0.88	0.91
Provolone	0.63	0.62	0.65	0.68	0.71
Parmesan	0.43	0.46	0.53	0.50	0.45
Romano	0.14	0.17	0.14	0.13	0.15
Other Italian	0.06	0.06	0.05	0.08	0.13
Miscellaneous cheeses	4.51	4.58	4.71	5.02	5.01
Cream	1.62	1.56	1.75	1.79	2.20
Swiss[b]	1.35	1.22	1.19	1.20	1.16
Muenster	0.40	0.42	0.45	0.45	0.44
Neufchâtel	0.10	0.21	0.27	0.31	NA
Blue[c]	0.17	0.16	0.15	0.15	0.16
Brick	0.07	0.06	0.06	0.05	0.05
Other miscellaneous cheeses	0.80	0.95	0.84	1.07	1.00
Processed cheese products	8.63	8.66	8.58	8.70	8.77
Processed cheese	4.79	4.89	5.23	5.23	5.29
Cheese foods and spreads	3.84	3.77	3.35	3.47	3.48
Total cheese[d]	24.60	24.80	25.90	26.20	27.00

Source: Dairy Management, Inc.

[a]Includes Colby, Monterey Jack, and washed curd and stirred curd.

[b]Includes Gruyère and Emmenthaler.

[c]Includes Gorgonzola.

[d]Total cheese excludes government removals. Individual cheese categories include government removals.

of $174 million the same year, the trend is toward expansion. The top cheese-importing countries in 1994 were Mexico (22 million pounds), Japan (4.7 million pounds), and Canada (4.2 million pounds). The growth potential for cheese exports from the United States looks bright due to passage of the North American Free Trade Agreement and the

Uruguay Round of GATT (General Agreement on Tariffs and Trade). As countries begin to phase down the use of export subsidies and other trade-distorting programs and as the economies of the world grow, the United States should face a strong world market for cheese and other dairy products.

Cheese Markets

The National Cheese Exchange (NCE) in Green Bay, Wisconsin, is, as the name implies, the national market for pricing cheese in the United States. It is a private, nonprofit corporation whose members buy and sell train carloads of cheese. The NCE was first established as the Wisconsin Cheese Exchange in Plymouth, Wisconsin, in 1918. While there were a number of cheese exchanges established over the years in other states—the first such exchange was established in 1871 in Little Falls, New York—the Wisconsin Cheese Exchange became the leading exchange as cheese production expanded in the state. The Wisconsin Cheese Exchange changed its name to the National Cheese Exchange and moved to Green Bay, Wisconsin, in 1975.

Very little cheese produced in the United States is actually traded on the NCE. In fact, Hamm and March (1995) report that over the period 1990–93 less than 1 percent of all cheese produced in the United States was actually traded on the NCE. The importance of the NCE, however, is that most cheese produced in the United States is contracted between buyer and seller at prices set relative to the NCE. It is customary to use formulas that price cheese at a premium or discount relative to the NCE based on the unique conditions and packaging of the cheese contracted for. Mozzarella cheese, for example, may be contracted between cooperative A and buyer B to be shredded and packaged at 5 cents above the NCE. The fact that so much of the nation's cheese is traded relative to posted prices on the NCE indicates the cheese industry's confidence that the exchange truly represents national supply and demand conditions. In general, prices traded on the NCE tend to rise when the milk supply tightens, demand for cheese is strong, and cheese stocks are down. On the other hand, prices fall when there is a surplus of milk on the market and cheese carry-in stocks are plentiful.

The NCE trades at 10:00 a.m. each Friday morning in Green Bay for about 30 minutes. Trading takes place in carlot units of cheddar cheese

(38,000–42,000 pounds). Units traded are in 40-pound blocks, 500-pound barrels, and 640-pound blocks. Trades are based on bids to buy and offers to sell. For example, a proprietary cheese plant may bid to buy four cars of 40-pound blocks at $1.3750 per pound. If accepted by a seller, the bid is filled. If it is too low, it is unfilled. On the other hand, an offer to sell may be accepted by a buyer, in which case the offer is covered. If the offer is too high, it is uncovered.

The National Cheese Exchange announced in March 1997 that it will be relocating to the Chicago Mercantile Exchange. The National Cheese Exchange came under intense public scrutiny after cheese prices in the fall of 1996 fell from record high levels of almost $1.70 per pound for 40-pound blocks to $1.19. Data, however, reveal that the exchange was merely reacting to supply and demand forces. The new exchange will use electronic trading via brokers to provide autonomous trading.

Another market that is emerging is the futures market for dairy products. Futures markets for milk and dairy products were not required back when these prices were set by support price announcements. Today, due to the overall reduction in the milk support price, tremendous volatility has occurred in both the milk and cheese market. This volatility will increase further if the U.S. dairy industry begins to export a significant quantity of its production onto the world market, or if the support price program and federal milk marketing orders are phased out or eliminated. The futures markets will emerge as an important tool for the dairy industry to manage price risk for dairy cooperatives, dairy producers, and milk handlers/processors.To date there are dairy futures contracts for fluid milk, butter, cheese, and nonfat dry milk offered by the Coffee, Sugar, and Cocoa Exchange in New York City and the Chicago Mercantile Exchange in Chicago.

Butter

Since the beginning of the U.S. dairy industry until the latter part of the nineteenth century butter was produced solely on the farm. Milk that was not immediately consumed was separated and churned by hand into butter. While butter was an important part of the farmer's diet, making it was a laborious task that was often performed by the farmer's wife. The usual routine was to milk the cows and set the milk in cellars or in cool springs until the cream rose. The cream was separated from the skim milk and then ripened by the bacteria present in the cream. After ripening the cream was churned by hand until the butter was formed. The churned butter was then worked and kneaded by hand to press out the buttermilk. The buttermilk that was left was either drunk as a farm delicacy or fed to livestock. The butter was then salted, packed in wooden crates, and stored in a cool cellar for later use.

Commercial butter production was initiated on the East Coast and centered in New York State. By the latter part of the 1800s, butter production shifted to the Upper Midwest and to the outer fringe of the Corn Belt. Selitzer (1976) reports that the top seven butter-producing states by 1910 were Wisconsin, Iowa, Minnesota, Pennsylvania, Michigan, Ohio, and Illinois. By this time butter factories, or creameries, began to expand. The first creamery built in the United States was in 1856 in Campbell Hall, Orange County, New York, by R.S. Woodhull. Farmers delivered their milk to the creamery where it was made into butter. Later, when Carl Gustaf DeLaval introduced the first practical cream separator in 1878, cream separation moved back to the farm. Dairy producers would separate the cream with DeLaval's hand-crank system and deliver just the cream to the creamery. Again, the liquid skim milk left after separation was fed to livestock, often to hogs. Farmers were separating milk on the farm with DeLaval's invention into the 1950s and 1960s when farm bulk tanks allowed farmers to have their milk picked up from the farm and delivered to the creamery.

The butter industry faced its most serious threat during World War II when per capita consumption began to fall. Per capita consumption of butter at the beginning of this century was about 20 pounds (Figure 5.1). America back then was largely an agrarian society, and butter, bread, meat, and potatoes were an important part of the farm diet. Butter was readily available since most farmers had at least one cow. It was also available in the cities since refrigeration wasn't necessary to store the product when handled properly.

The reason for the decline in butter consumption was due in part to greater consumption of oleomargarine, a nondairy table spread. Oleomargarine was invented in 1867 by the French chemist Hippolyte Mège-Mouriéz, who entered a contest sponsored by Napoleon III. The contest awarded a prize for anyone who found a satisfactory butter substitute to be used by the navy and the poor (Riepma 1970; Selitzer 1976). Mège-Mouriéz used margaric acid, a fatty acid component he derived from finely minced beef. Margaric acid was in fact first isolated by

Michael Eugène Chevreul in 1813. Chevreul named the product "margaric" because the lustrous pearly drops of the product reminded him of the Greek word for pearls—*margarites*. Mège-Mouriéz knew he needed to find a name for his product in order to differentiate it from butter. He came up with the word "oleomargarine" because so much of his product consisted of margaric acid. The prefix "oleo-" was taken from the Latin word *oleum*, which is a name for beef fat, the principal ingredient used. The word "oleomargarine" was later simplified to "margarine."

Margarine was first introduced in the United States when a Mr. Paraf opened the Oleo-Margarine Manufacturing Company in New York City in 1873. The product was not an immediate success since it was considered a poor man's butter. Even though its price was much cheaper than that of butter, it was considered an inferior product. In addition, new laws were enacted to restrict consumption of margarine. New York was the first state to enact an outright ban on the use of the product in 1884. This law was struck down by the courts a year later. Congress then got into the act and passed the Margarine Act of 1886, which imposed a tax of 2 cents a pound on margarine and required manufacturers and dealers to be regulated by licenses. Antimargarine laws were amended and enacted at the state and federal levels over the intervening years.

Margarine consumption didn't really expand until World War II, when butter was relatively scarce due to rationing and margarine was

Figure 5.1. Per capita consumption of margarine and butter

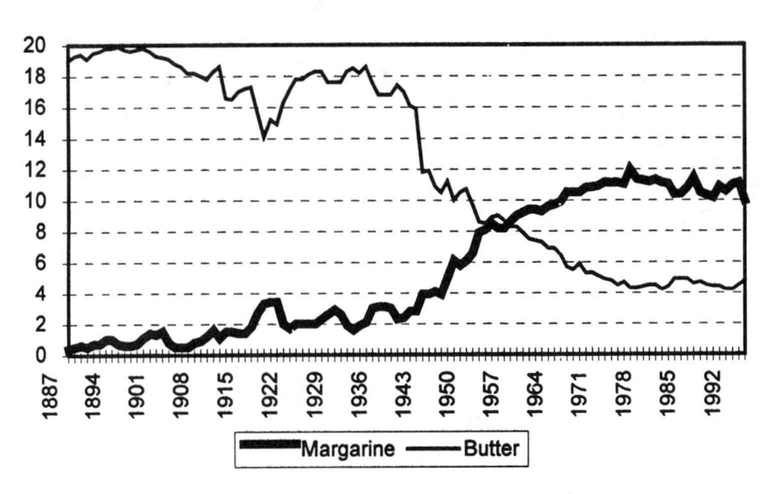

plentiful. As a result, per capita consumption of butter declined from 17 pounds in 1940 to 10 pounds in 1948. At the same time, per capita consumption of margarine rose from 2.4 to 6.1 pounds. Selitzer (1976) reports that butter was then dealt a triple blow in 1950: the cholesterol controversy flared up, the U.S. Congress acted to remove all special taxes against margarine, and the price of butter skyrocketed to almost twice the price of margarine. By 1975 per capita consumption of butter fell to just 4 pounds while per capita consumption of margarine rose to 11 pounds.

Standards and Classifications

Butter is graded by the USDA, which requires rigorous production standards be met in order to bear the USDA grade shield. There are presently four grades of butter: Grade AA, Grade A, Grade B, and Grade C. The first national grading system for butter was developed by the USDA in 1918. The USDA published uniform standards to be followed by government graders in all markets. Then in 1924 Land O'Lakes Creamery petitioned the USDA to set up grading at the point of production for its St. Paul and Duluth warehouses. This was a bold move since butter was never graded at the plant site. But the cooperative built its reputation on the basis of quality and wanted its butter to bear the seal of approval of federal inspectors. The grade shield was an immediate hit with consumers, and federal grading at the point of production eventually became a national standard.

Today all butter is graded by the USDA at the plant where it is manufactured or packaged. To bear the USDA grade shield butter must be churned from pasteurized cream in an approved plant under sanitary conditions (Campbell and Marshall 1975). Grades are established based on characteristics of flavor, body, color, and salt. Grades are assigned based on a possible total score of 93. USDA grades of AA, A, B, and C correspond with numerical scores of 93, 92, 90, and 89, respectively.

Butter Processing

Butter today is processed from fresh sweet cream. As stated earlier, this was not always the case. At the turn of the century it was customary to allow cream to sour before it was churned. The microorganisms in the milk formed an acid that resulted in a very strong flavor in the butter. At the time it was thought that the acid aided in the storability of butter. In fact, it was later discovered that just the opposite was true. Today, only fresh sweet cream is used in the production of butter.

Campbell and Marshall (1975) outline the process for making butter, which begins with the clarification and separation of milk. Cream with a concentration of 30 to 45 percent milkfat (depending on the method of churning) is then pasteurized and cooled. For vat pasteurization the cream is normally pasteurized at 165°F (74°C) for 30 minutes; for the high-temperature short-time method cream is pasteurized at 185°F (85°C) for 15 seconds. These pasteurization temperatures are higher than those for fluid milk (1) because of the higher fat content of the cream and (2) to help lengthen butter's shelf life. The cream is not homogenized since that would make churning more difficult.

After the cream is cooled, it is pumped into a conventional churn where it may be mixed with anotto yellow coloring. The cream is then churned until butter granules are formed. The buttermilk is drained and washed from the butter granules, salt is added, and the butter is worked to a smooth, creamy consistency. The butter is then packaged by a print machine, which molds it into sticks, wraps it, and packages it. The term "printing" came from the days when butter was pressed into a mold and the manufacturer's label was printed on the butter. Butter is also packaged in many other size containers including 60-pound boxes for sale to the Commodity Credit Corporation (CCC) under the price support program, as well as in chips, patties, continentals, and reddies for sale to the consumer.

Anhydrous milkfat and anhydrous butter oil are two butter products that are high in fat content and very low in moisture (Sadler and Wong 1970). Anhydrous milkfat is made from fresh whole milk. Cream containing 40 percent milkfat is first separated from milk. It is then reconcentrated to 75–85 percent milkfat by heating the cream to 130°–150°F (54°–66°C) and using centrifugal separation. Then using a process called "phase inversion," the cream is prepared for dehydration. The subsequent crude oil derived from phase inversion is then further concentrated in a centrifugal separator to 96–98 percent oil. This oil is then heated to 155°–190°F (68°–88°C) and spray injected into a vacuum chamber to remove moisture. This final product is not less than 99.8 percent fat and has less than 0.1 percent moisture. Anhydrous butter oil, on the other hand, is manufactured directly from butter. Butter is converted to oil by melting the butter, washing it with water, and then dehydrating it. The final product is similar in content to anhydrous milkfat. Anhydrous milkfat and anhydrous butter oil are two products that are commonly traded in the international market.

The "Butterfat Problem"

The decline in the per capita consumption of butter created a serious butterfat problem in the early 1980s and again in the early 1990s. Government removals of butter from the market under the Dairy Price Support Program exceeded 400 million pounds per year (Figure 5.2). This added significantly to the cost of the program. Part of the problem was the economic incentive supplied by high support prices to produce butter and to offer it for sale to the government. As a result, Congress and a number of presidential administrations began a careful plan to phase down the CCC purchase price for butter from a high of 153 cents per pound on April 1, 1981 (for Grade A butter at Chicago), to 65 cents per pound on July 7, 1993 (Blayney et al. 1995).

Figure 5.2. Comparison of government removals and the wholesale price of butter

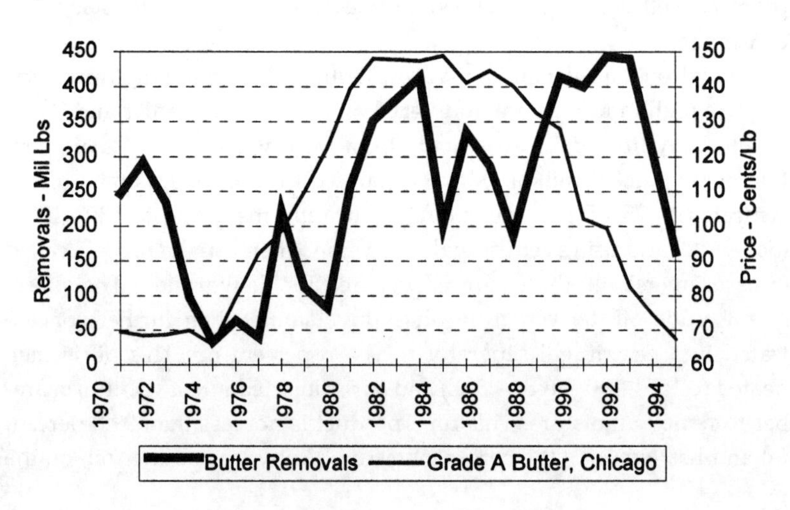

The results have been dramatic. The lower support price for butter has lowered its market price and the butterfat premium paid to dairy farmers. As a result, less butter has been offered for sale to the CCC. In addition, butter consumption increased 16 percent in 1994 over the previous year due to three factors: (1) lower butter prices in comparison with margarine prices, (2) renewed consumer interest in butter as a "natural product" with unmatched flavor, and (3) increased export demand for U.S. butter. A world shortage of butter in 1995 resulted in significant exports of butter from the United States with no Dairy Export Incentive Program (DEIP) subsidies. As a result, CCC uncommitted inventories of butter declined from 92.8 million pounds on September 8, 1994, to zero on September 8, 1995. In fact, the USDA had to purchase butter from the commercial market during the summer of 1995 in order to meet the needs of domestic feeding programs.

In retrospect, the "butterfat problem" was nothing more than a problem created by a price-distorting government support program. When the support price of butter was phased down, market-clearing prices were discovered. Then, when excess government supplies were exhausted and domestic prices became competitive with international prices, the United States became a butter exporter. In 1995 butter prices rose from 65 cents per pound in January to $1 by the fall as the market faced a significant shortage. (The lessons to be learned here are discussed in more detail in Chapter 11.)

Dried Dairy Products

Two major problems facing the dairy industry since its inception were (1) how to store milk for later use and (2) how to make use of dairy byproducts left over from manufactured products like cheese and butter. The earliest records of dry milk were written by Marco Polo in the thirteenth century when he observed the use of sun-dried milk by the Mongolians (Selitzer 1976). The technology to properly remove moisture from dairy products wasn't developed till the latter part of the nineteenth century. The first patent for a drying machine was issued in 1899 to William Gere and Irving Merrell. The first successful spray drying plant was started in 1905 in Arcade, New York. Significant refinements have since been made to improve the economies of production and the flavor of dried dairy products. The result has been the creation of a number of dairy products with many different uses and a more economical recovery of by-products.

The most common dried milk products produced and consumed in the United States are nonfat dry milk (called "skim milk" powder in international circles), casein, whey and modified whey products, lactose, and dried cream, butter, and cheese. While some of these products (i.e., casein, whey, lactose, and perhaps nonfat dry milk) are clearly by-products, others are not (dried cream, butter, and cheese). Most of these products are used for further processing of other dairy products, for home use, as an ingredient in the baking and confectionery industry, as a livestock feed, in meat processing, and in the chemical and pharmaceutical industries.

Most dairy products are dried in one of two ways: (1) by drum or roller driers or (2) by spray or foam spray drying. The key to successfully removing moisture from dairy products is to not affect heat-sensitive milk components (Webb and Whittier 1970). That's why lower-cost methods such as boiling are not used. Proteins in milk, for example, can easily be destroyed by heat, and the flavor of the product adversely affected. Drum or roller drying applies a thin layer of milk or by-product concentrate onto the surface of a rotating metal drum that is heated from the inside with steam. The thin film then dries as the drum rotates and is scraped off by stationary steel blades. For small volumes drum drying requires less space and is more economical than other methods of drying (Hall and Hedrick 1971). But drum or roller dryers are seldom used today because the high temperatures render milk proteins insoluble, can adversely affect color, and may produce a scorched flavor. With spray driers the product is first condensed in a vacuum pan or evaporator and then atomized and sprayed into a stream of hot air. A collection system used in a large chamber separates the dried particles from the moist air.

Nonfat Dry Milk

Nonfat dry milk (NDM) is produced from pasteurized skim milk. Two methods of NDM production are low-heat and high-heat. Low-heat NDM requires that the heating used during pasteurization and drying be carefully controlled in order to avoid denaturing the whey protein. Campbell and Marshall (1975) note that low-heat NDM is preferred for table use and for cakes, custards, and other similar products. The whey proteins are denatured in high-heat NDM, which is preferred by bakers because the loaf volume of baked products is improved and it retains moisture better than low-heat NDM.

The *CFR* defines NDM based on the drying process. NDM is the product resulting from the removal of fat and water from pasteurized milk and contains the lactose, milk proteins, and milk minerals in the same relative proportions as the fresh milk from which it is made. It contains not over 5 percent by weight of moisture. The USDA recognizes two grades of NDM: Extra and Standard. These grades are determined on the basis of flavor, physical appearance, bacterial estimate on the basis of standard plate count, milkfat content, moisture content, scorched particle content, solubility index, and titratable acidity. In addition to grades, *CFR* has a heat treatment classification (for spray process NDM) that processors may request USDA to show on the grading certificate. Under that grading system, U.S. high-heat is NDM that does not exceed 1.5 milligrams of undenatured whey protein nitrogen per gram of NDM. U.S. low-heat does not have less than 6.0 milligrams of undenatured whey protein nitrogen per gram of NDM. And U.S. medium-heat has undenatured whey protein nitrogen between the levels of high-heat and low-heat.

The most typical uses of NDM is for further processing into dairy and bakery products and for instant NDM for home use. NDM is used in the processing of many dairy products such as fluid milk, ice cream, cheese, and yogurt. NDM is also used in bread dough formulas since it improves flavor, texture, and grain and crust color and improves water absorption capacity (Webb and Whittier 1970). It is also used in processed meats to improve their water-binding capacity. In 1993, 64 percent of all domestic use of NDM was for dairy use, 11 percent for bakery use, and 10 percent for home use (Table 5.3).

The production of NDM has declined from a high of 1.5 billion pounds in 1983 to 945 million in 1993 (Table 5.4). Most of this decline has been associated with changes in demand. Per capita consumption of NDM declined from 3.4 pounds in 1976 to 1.8 pounds in 1985 (Figure 5.3). Thereafter it rose to 3.5 pounds in 1989, only to fall to 2.4 pounds by 1994. The consumption of NDM is sensitive to the price of the product, the price of substitutes, and government programs such as the Dairy Price Support Program and the DEIP. In recent years the CCC purchase price of NDM was raised relative to that of butter in order to lower the

Table 5.3. Domestic end uses of nonfat dry milk in 1993

	Million lb	Percentage of domestic uses[a]
Dairy (reprocessing into other products)	360.2	64
Bakery	61.9	11
Packaged for home use	59.1	10
Prepared dry mixes	11.3	2
Chemicals/pharmaceuticals	11.3	2
Meat processing	13.5	2
Other[b]	45.6	8
Total domestic nongovernment use	562.9	100

Source: Milk Industry Foundation 1994.

[a]Figures rounded off.

[b]Includes use in confectionery, soft drink bottling, and soup and margarine. Also includes use in institutions and as animal feed.

overall cost of the Dairy Price Support Program. As a result, the market price of NDM rose. Also, the introduction of Class IIIa pricing created an artificial incentive to overproduce NDM, which has resulted in CCC accumulation of surplus NDM. In addition, the DEIP program has been aggressively used to market surplus NDM. All of these programs have acted to distort the true market price of NDM.

Dry Whole Milk

Dry whole milk is very similar to NDM with the exception that milkfat is retained in the product. The *CFR* defines dry whole milk as the product obtained by removal of water from pasteurized milk that may have been homogenized. The product should contain the lactose, milk proteins, milkfat, and milk minerals in the same relative proportions as the milk from which it is made. It may be fortified with either vitamins A or D or both. Two grades are recognized by the USDA: Extra and Standard. The single largest use of dry whole milk is in the candy industry. Milkfat has a fairly low melting point (82°–99°F or 28°–37°C) and

Table 5.4. Production of nonfat dry milk, whey, and modified whey products

Year	Dry whey[a]	Condensed sweet whey[a]	Condensed acid type[a]	Modified dry whey products: Part delactosed and demineralized	Whey protein concentrate	Lactose	Whey solids in wet blends	Total whey products[b]	Nonfat dry milk[c]
				(thousand lb)					
1980	689,718	81,379	4,322	192,878	NA	140,177	144,066	1,252,540	1,160,691
1981	778,720	89,936	23,026	173,545	NA	162,090	160,375	1,387,692	1,314,270
1982	790,502	119,674	21,978	234,175	NA	144,356	136,325	1,447,010	1,400,455
1983	892,164	117,341	19,163	115,968	86,308	130,335	123,980	1,485,259	1,499,902
1984	897,946	119,597	10,425	87,769	96,298	123,778	135,692	1,471,505	1,160,670
1985	986,829	116,291	10,380	96,355	104,683	123,173	136,341	1,574,052	1,390,033
1986	1,031,033	109,063	8,797	90,401	78,049	135,181	116,410	1,568,934	1,059,049
1987	1,097,349	67,878	5,476	106,964	97,353	151,522	134,224	1,660,766	1,056,797
1988	1,136,987	57,908	4,217	121,733	135,860	161,296	128,308	1,746,309	979,722
1989	1,069,470	59,889	30	114,995	142,216	170,903	108,894	1,666,397	874,667
1990	1,143,259	58,683	0	94,871	168,056	192,586	107,124	1,764,579	879,212
1991	1,167,394	33,415	0	107,762	185,048	196,879	102,997	1,793,495	877,525
1992	1,237,283	41,416	0	118,894	178,361	246,873	118,503	1,941,330	872,123
1993	1,196,378	49,698	181	109,457	174,181	236,381	113,352	1,879,628	948,117
1994	1,211,787	85,404	0	95,820	181,813	253,261	99,808	1,927,893	1,230,855

Source: National Cheese Institute 1995.

Note: NA = not available.

[a]Final marketable product only. Does not include quantity used or shipped to another plant for further processing into dry whey or modified dry whey products.

[b]For human food and animal feed.

[c]For human food.

can be easily blended with other fats in order to help give candy a firm but pliable body. In 1993 88 percent of all dry whole milk was used by confectioneries (Table 5.5).

Dry whole milk is also used in processed dairy products, bakery goods, and the production of other prepared foods. It is also a product that is frequently exported since it does not need the addition of milkfat when rehydrated.

Dry Buttermilk

Dry buttermilk is essentially a by-product of the manufacture of sweet cream butter (Webb and Whittier 1970). Dry buttermilk is used primarily in the dairy processing and the bakery industries. In 1993, 36 percent of all dry buttermilk was used in dairy products and 23 percent in bakery goods (Table 5.6). High-acid dry buttermilk made from sour cream butter contributes to a distinctive flavor in rye and whole wheat breads. Dry buttermilk also improves the whipping properties and flavor of various frozen desserts. Its use can also improve the texture and stability of recombined dairy products.

The *CFR* defines dry buttermilk as the product resulting from drying liquid buttermilk that was derived from the churning of butter and pasteurized prior to condensing at a temperature of 161°F (72°C) for 15 sec-

Figure 5.3. Per capita consumption of select dried dairy products

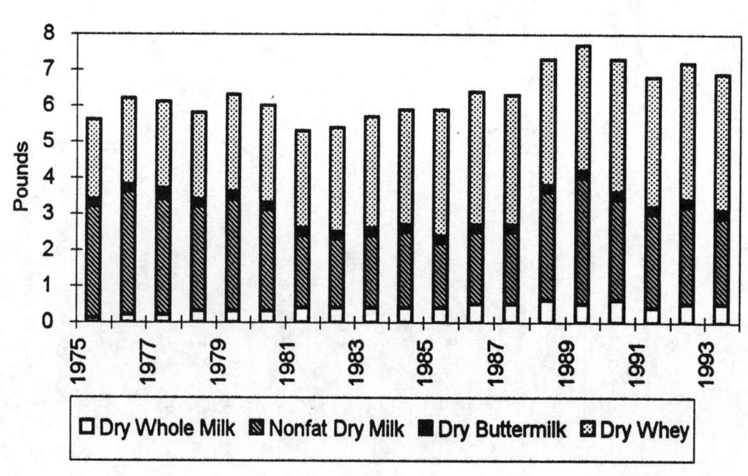

Table 5.5. Domestic end uses of dry whole milk in 1993

	Million lb	Percentage of domestic uses
Confectionery	85.5	88
Dairy (reprocessing into other products)	4.2	4
Bakery	4.4	5
Prepared food manufacturers	2.0	2
Other[a]	1.1	1
Total domestic sales	97.2	100

Source: Milk Industry Foundation 1994.

[a]Includes all other domestic uses.

Table 5.6. Domestic end uses of dry buttermilk in 1993

	Million lb	Percentage of domestic uses
Dairy (reprocessing into other products)	31.6	36
Bakery	19.7	23
Prepared dry mixes	13.0	15
Confectionery	12.9	15
Other[a]	9.8	11
Total domestic sales	87.0	100

Source: Milk Industry Foundation 1994.

[a]Includes all other domestic uses.

onds for proper bacterial destruction. Dry buttermilk must have a protein content of not less than 30 percent and should not contain or be derived from nonfat dry milk, dry whey, or products other than buttermilk. It should contain not less than 4.5 percent milkfat and not more than 4 percent moisture. Dry buttermilk should not contain any added preservative, neutralizing agent, or other chemical. USDA-approved grades for dry buttermilk are Extra and Standard.

Dry buttermilk accounts for a very small fraction of production and sales of dried dairy products. For example, in 1994 the United States produced 1.44 billion pounds of dry milk products. Of that amount 84 percent was NDM and just 3.6 percent by weight was dry buttermilk. This was due primarily to low per capita consumption (Figure 5.3).

Dry Whey and Whey Products

The major by-product of cheese production is whey. Campbell and Marshall (1975) note that about half the original milk solids are left in whey from the manufacture of most cheeses. The average composition of whey from American cheese is 4.9 percent lactose, 0.9 percent protein (includes 0.1 percent casein), 0.6 percent ash, 0.3 percent fat, and 0.2 percent lactic acid. Lactose is the sugar that is naturally present in milk. Casein is a protein that occurs naturally in milk and is a major ingredient in cheese. The whey from rennet-set cheeses is a sweet whey since its pH is approximately 6.0. As such it is more easily dehydrated than sour or acid whey from cottage cheese, which has a pH of about 4.6. In the past, whey from large cheese processing facilities was thrown away, fed to livestock, and/or land-applied as a fertilizer. This not only resulted in a waste of valuable nutrients but also caused significant environmental problems. As a result, numerous amounts of research led to methods for harvesting the nutrients from whey either in dry or wet form. As a result the production of whey products today exceeds that of nonfat dry milk (Table 5.4).

One way to process whey from cheese production is to dry it. Sweet dried whey's processing is similar to NDM's: it is pasteurized, condensed, dried, and packaged. One problem encountered in spray drying whey, however, is that, due to its lactose content, it can become hygroscopic (i.e., lactose readily absorbs water, which results in caking and loss of flow properties). New methods are used today to produce a nonhygroscopic product.

A major use of dried whey is as a poultry and livestock feed (Webb and Whittier 1970). Other uses are as an ingredient in dairy products, dry blends, prepared dry mixes, bakery products, soup, and confectionery products. The *CFR* defines dry whey as the product resulting from drying fresh whey that has been pasteurized and to which nothing has been added as a preservative. Dry whey contains all of the constituents, except moisture, in the same relative proportions as in the whey. Campbell and Marshall (1975) note that dried whey contains approximately 70 percent lactose, 12 percent protein, and 8 to 10 percent ash. Dried whey should contain not more than 1.5 percent milkfat and not more than 5 percent moisture. There is only one grade of dry whey: Extra. This grading is determined on the basis of flavor, physical appearance, bacterial analysis, titratable acidity, milkfat content, moisture, and color.

There are many other products that can be processed from liquid whey. These include lactose, casein, and modified whey products such as whey protein concentrate and partial delactosed and demineralized whey. Lactose has historically been used in infant formulas because of its ease of digestibility. It is also used as a flavor enhancer in many foods such as flavored potato chips, barbecue sauce, and salad dressings. Besides use in pharmaceuticals, lactose is also used in caramels and fudges (to improve body, texture, chewyness, and shelf life) and in bakery products. Casein has largely been imported into the United States since it has been cheaper to import the product than to process it. Major uses of casein are (1) paper coating, (2) adhesives, (3) foods, (4) plastics, and (5) man-made fibers (Webb and Whittier 1970). The major use of casein in food production is in the processing of artificial dairy products such as artificial cheese, coffee whiteners, whipping powders, instant breakfasts, and imitation milks. Whey proteins can also be separated from whey and used as a component in bakery goods, baby foods, dry pudding mixes, ice cream mixes, salad dressings, and whipping agents and as an emulsifying agent for toppings. It can also be used in soups and bouillon, fruit juices, condiments, and dietetic preparations.

NOTES

1. "Cheeses and Related Cheese Products," *Code of Federal Regulation*, title 21, pt. 133, 1995 ed.
2. "United States Standards for Grades of Bulk American Cheese," *Code of Federal Regulation*, title 7, pt. 58, subpt. H, 1995 ed.
3. The exceptions are cottage cheese and cream cheese, which take longer, about 4–16 hours.

REFERENCES

Battistotti, Bruno, Vittorio Bottazzi, Antonio Piccinardi, and Giancarlo Volpato. 1984. *Cheese: A Guide to the World of Cheese and Cheese Making*. New York: Facts on File, Inc.

Blayney, Don P., James J. Miller, and Richard P. Stillman. 1995. *Dairy: Background for 1995 Farm Legislation*. Economic Report 705. Commercial Agriculture Division, Economic Research Service, U.S. Department of Agriculture, Washington, D.C., April.

Campbell, John R. and Robert T. Marshall. 1975. *The Science of Providing Milk for Man*. New York: McGraw-Hill Publishing Co.

Eekhof-Stork, Nancy. 1976. *The World Atlas of Cheese*. Edited by Adrian Bailey. New York: Paddington Press.

Hall, Carl W. and T.I. Hedrick. 1971. *Drying of Milk and Milk Products*. Second ed. Westport: AVI Publishing Company.

Hamm, Larry G. and Robert March. 1995. *The National Cheese Exchange: Impacts on Dairy Industry Pricing*. M-7. Dairy Markets and Policy Issues and Options. Cornell University, February.

Milk Industry Foundation. 1994. *Milk Facts: 1994 Edition*. Washington, D.C., September.

National Cheese Institute. 1995. *Cheese Facts: 1995 Edition*. Washington, D.C., October.

Riepma, S.F. 1970. *The Story of Margarine*. Washington, D.C.: Public Affairs Press.

Sadler, A.M. and N.P. Wong. 1970. "Milk Fat Utilization in Foods." In *By-products from Milk,* second ed. Westport: AVI Publishing Company.

Scott, R. 1986. *Cheesemaking Practice*. Second ed. London and New York: Elsevier Applied Science Publishers.

Selitzer, Ralph. 1976. *The Dairy Industry in America*. New York: Dairy Field and Ice Cream. Books for Industry, Divisions of Magazines for Industry.

U.S. Department of Agriculture. National Agricultural Statistics Service. 1995. *Dairy Products: 1994 Summary*. Da 2-1(95). Washington, D.C., May.

Webb, Byron H. and Earle O. Whittier. 1970. *Byproducts from Milk*. Westport: AVI Publishing Company.

Part II

Government Policies and Regulations on Milk Marketing

Chapter 6

Federal Milk Marketing Orders

One of the most difficult lessons for any serious student of milk marketing is to learn about U.S. federal milk marketing orders (FMMOs). FMMOs are a complex set of rules that assist in the equitable pricing of milk to dairy producers and processors and help provide an adequate supply of pure and wholesome milk to consumers. They were developed to enhance prices to dairy producers and to help create orderly marketing conditions. But the rules and requirements used to operate federal orders are mind-boggling to the beginner, often prompting the question, "Isn't there a simpler way to do this?"

To proponents of FMMOs, they may be analogous to traffic rules and regulations that were designed to assist travelers and promote the orderly flow of traffic. While eliminating such rules may be ideologically appealing, an everybody-for-themselves mentality would likely prevail, resulting in chaos and confusion that would disrupt traffic flow.

The reason FMMOs are complex is basically due to two realities. First, milk is a highly perishable commodity that must be produced, transported, processed, and marketed under strict sanitary conditions. It cannot be stored in fluid form for very long and must be sold to the consumer within one or two days after it is harvested from the cow. Dairy producers cannot withhold milk from the market in order to wait for prices to improve. Also, dairy producers must have a reliable market for their milk 365 days a year since a cow cannot be turned on and off like a faucet. Therefore, by its very nature, the marketing of milk is very different from that of other storable commodities or products.

Second, FMMOs represent government intervention in the marketplace in order to offset otherwise chaotic marketing conditions. Some would argue that without FMMOs milk prices would swing wildly between surplus and shortage periods of the year, processors would be unwilling to pay farmers a higher price for milk used for bottling purposes, and consumers would pay more as many smaller producers and processors would be forced to leave a competitive market environment. FMMOs essentially amount to administering the way milk is priced and marketed according to the public interest in an attempt to promote the orderly marketing of milk in a fair and equitable way. Therefore, rules must be clearly stated in order to represent the interests of dairy producers, processors, and consumers.

This concept of public interest and its relation to the marketing of milk was discussed in depth in a report issued in 1962 by a blue-ribbon committee formed by Secretary of Agriculture Orville Freeman and chaired by Edwin G. Nourse (Federal Milk Order Study Committee 1962). The committee interpreted the public interest criterion applied to the order system by the original implementing legislation for FMMOs. This required the secretary of agriculture to bring producers, consumers, and processors into a desirable commercial equilibrium that could be obtained through optimum allocation of the nation's resources. The goal was an administered pricing structure that meets the needs of all three groups. But it is just this process of replacing competitive market forces with an administered pricing structure that makes FMMOs so complex.

This chapter provides a detailed overview of FMMOs starting with the reasons for the original legislation back in the early 1930s. A detailed discussion of how the system works is presented in order to provide the serious student with a solid foundation in the economics of federal orders and an understanding of how these orders affect milk marketing in the United States. It is essential to learn this information in order to comprehend and contribute to current proposals to refine and improve the federal order system.

Background and History

Federal orders have a long and interesting history. They have their origins in the chaotic marketing conditions that occurred in the early 1930s as a result of the Depression. Low milk prices, due to the loss of purchasing power of consumers and a near collapse of the economy, prompted Congress to enact emergency legislation that resulted in the creation of FMMOs.

During the 1920s, dairy farmers did fairly well compared with other agricultural producers in terms of income. According to Manchester (1983), cash receipts from marketing dairy products rose 31 percent between 1924 and 1929 compared with just a 7 percent gain for the rest of agriculture. This was due in large part to cooperatives that provided dairy farmers greater market power in negotiating prices with processors and milk dealers. Cooperatives developed classified pricing plans, which established a schedule of milk prices depending on how milk was used by processors. Such plans segregated milk into two uses: one for bottled milk (for fluid uses) and the rest (considered surplus milk) for manufacturing into storable dairy products such as butter, nonfat dry milk, and cheese. Classified pricing plans placed a higher value on milk used for fluid uses and a lower value on milk used to manufacture dairy products.

The other reason that milk prices were higher during this period was because fluid processors were willing to pay cooperatives a higher price for their milk since classified pricing plans prevented other dealers from starting a price war in the fluid market (Justice Department 1977).

This system began to collapse, however, when milk prices fell in response to the Depression. The average price of milk dropped 31 percent between 1929 and 1932 (Manchester 1983). This was due in large part to the reduced purchasing power of consumers. In addition, some milk dealers that specialized in fluid milk sales found it profitable to purchase milk from dairy producers who did not belong to a dairy cooperative. They offered these farmers a price that was lower than what they were required to pay the cooperative, but higher than what the cooperatives could pay their members. Milk dealers could do this since classified pricing plans in those days were strictly voluntary. As a result, cooperatives did not have the authority to audit the books of processors in order to ensure that their members were being paid the higher price for milk used for fluid purposes, which likely resulted in widespread underpayment to dairy producers.

Farmer organizations and some cooperatives responded by forming milk strikes to try to curtail national production and to raise milk prices. Violence erupted in many cities where outside milk entered a market that was embargoed by striking dairy farmers (McMenamin and McNamara 1980). These strikes, however, had little impact on production and milk sales to processors.

Congress responded to this situation with emergency measures under the Agricultural Adjustment Act (AAA) of 1933 (P.L. 73-10, May 12, 1933). The bill was enacted as a stopgap measure during a period of extreme crisis. The primary purpose of the legislation was to improve the agricultural purchasing power of farmers by raising their income. The major features of this legislation were processing taxes, an allotment plan, and licenses and marketing agreements. Taxes on processors were used by the government to lease land from farmers, thus reducing supplies. The allotment plan allowed the government to pay farmers parity prices on their domestic allotted production. Parity was a concept used to raise the farm price of agricultural commodities to restore the purchasing power of farmers to the level that occurred during the five-year period immediately preceding World War I (Nicholson 1965). In fact a major objective of the act was to "reestablish prices to farmers at a level that will give agricultural commodities a purchasing power with respect to articles that farmers buy, equivalent to the purchasing power of agricultural commodities in the base period ... August 1909–July 1914" (USDA, AAA 1933).

Section 2 of the act provided the secretary of agriculture with broad powers to establish marketing agreements and to enforce such agreements with licenses. The marketing agreement was to be a voluntary agreement entered into by the secretary of agriculture with processors, associations of producers (cooperatives), and other handlers of milk (Hutt 1960). Congress also intended the licenses to prevent competitive price cutting by milk distributors, which at the time was considered an unfair trade practice (Justice Department 1977). The AAA of 1933 thus required milk distributors and processors to pay minimum prices to dairy producers.

The AAA of 1933 did not spell out specific provisions for these marketing agreements and licenses—how they were to operate and how they were to further the intent of the act. In fact, there was no language that clearly stated how milk was to be priced other than an overall objective of having farm prices achieve parity. It was left to the secretary of agriculture, who had broad powers to interpret and administer the act.

The first milk marketing agreement and license was for the Chicago market and was effective on August 1, 1933. By August of the following year, a total of 50 markets were issued licenses (Nicholson 1965). One such marketing agreement and license was announced by the secretary of agriculture on November 22, 1933, for milk in the St. Louis market

(USDA, AAA 1933). The terms of the agreement were administered by a Milk Industry Board. That agreement set up a schedule governing the prices and terms under which milk was to be sold. The price schedule was for both wholesale and retail prices for all dairy products sold in the St. Louis market. The license for milk in the St. Louis market allowed the secretary of agriculture to license and regulate every distributor of fluid milk in the St. Louis sales area. In addition, it required that milk could be sold to consumers only if it complied with standards governing the production, receiving, transportation, processing, and distribution of fluid milk as set by the health ordinances of St. Louis.

These first marketing agreements and licenses set prices via a classification use plan that had already been used in the industry since 1916 (Hutt 1960, n. 28, p. 532). This plan was based on two concepts: classified pricing and marketwide pooling.

Classified pricing allowed milk to be valued according to its ultimate use. The Class I price was for milk used for bottling purposes. It was the highest price since milk used for bottling purposes (Grade A milk) had to meet tough sanitary regulations that were already enacted at the time by many states and municipalities. Those regulations affected dairy producers by raising the cost of producing Grade A milk well above that for manufacturing grade milk. The other classes were for lower-valued surplus milk, or Grade A milk in excess of fluid needs.

Marketwide pooling required all processors in a given market to place the proceeds of classified milk sales into a single pool from which an average price, the blend price, was calculated each month by dividing total proceeds by the pounds of milk sold in the marketing order. Marketwide pooling allowed all members of the pool to share equitably in both the higher returns from the fluid market as well as the lower returns from the surplus market. Thus a single price was paid to all producers in the marketing agreement. This eliminated the potential situation where one producer, who sold milk to a fluid processor, would get a higher milk price than his neighbor who sold to a cheese plant.

The AAA was later amended by the Agricultural Adjustment Act of 1935 (P.L. 74-320, August 24, 1935). A Supreme Court decision in the *Schechter Poulty* case led to doubt over the constitutionality of the AAA of 1933. At question was the legality of the secretary's action in certain areas of regulation (Nicholson 1965). The amended legislation replaced licenses with orders and further refined the broad powers of the secretary of agriculture. Title I, Section 8c, of the AAA of 1935 specifically granted the secretary of agriculture with powers to issue and amend "orders, applicable to processors, associations of producers (cooperatives), and others engaged in the handling of any agricultural commodity" (USDA, AMS 1989). In addition, the amendment continued the formula for parity prices using the base period August 1909 to July 1914 (Justice Department 1977). "Parity prices" refers to a measure of how well prices for agricultural commodities during a specified time period compare with prices during the period 1910–14, when the relationship between prices received for agricultural commodities and prices paid for production goods and services was favorable to agricultural producers and manufacturers.

The AAA of 1935 specifically authorized the major provisions of federal orders that are still in effect today, including classified pricing and pooling. Other provisions included checking weights and tests, enforcing base and surplus plans, requiring proposed federal orders to have the approval of two-thirds of producers voting in a referendum for marketwide pools, and the right of cooperatives to vote on behalf of their members (Nicholson 1965).

The first two federal orders issued by the secretary of agriculture were for the Kansas City and St. Louis marketing areas and became effective February 1, 1936. These orders laid very specific guidelines for administering the Kansas City and St. Louis federal orders. The parity pricing concept was used in setting class prices, and the order changed the parity base period to January 1923 through July 1929 in order to establish the purchasing power of milk. In addition, two classes of milk and minimum prices for each were established. These prices were to be blended back to producers via a marketwide pool. A market administrator was to be selected by the secretary of agriculture to administer the order. In addition, necessary expenses for operating the orders were set. Milk dealers were to periodically report to the market administrator receipts of milk purchased and quantities of milk that were sold or processed for the purpose of classification. Many of these same principles are still used in today's federal orders.

The Agricultural Marketing Agreement Act (AMAA) of 1937 (P.L. 75-137, June 3, 1937) further modified the AAA of 1935 to provide a framework for long-run price and marketing stability (USDA, AMS 1989). Debate at the time centered on whether Congress had the constitutional power to regulate the interstate shipment of and international trade in agricultural products through marketing orders as provided by the AAA of 1935.

The AMAA of 1937 established criteria for the secretary of agriculture to establish minimum prices in federal orders. Congress recognized that parity pricing was not an appropriate method of establishing class prices under federal orders (USDA 1972). Under Section 8(c)18 of the amended legislation, if the national parity price for milk as defined under the AAA of 1935 was not reasonable in any particular order, the secretary had the authority to set producer prices in that order to ensure a sufficient quantity of milk. In particular, the act states,

> Whenever the Secretary finds ... that the parity prices of such commodities (milk or its products) are not reasonable in view of the price of feeds, the available supplies of feeds, and other economic conditions which affect market supply and demand for milk and its products in the marketing area to which the contemplated agreement, order, or amendment relates, he shall fix such prices as he finds will reflect such factors, insure a sufficient quantity of pure and wholesome milk to meet current needs and further to assure a level of farm income adequate to maintain productive capacity sufficient to meet anticipated future needs. (USDA, AMS 1990)

The major intent of Congress in enacting the AMAA of 1937 and previous legislation was "to establish and maintain such orderly marketing conditions for agricultural commodities in interstate commerce as will establish, as the price to farmers, parity prices" (AAEA Task Force on Dairy Marketing 1986, p. 3). Manchester (1983) defines orderly marketing as "spreading the marketing of a farm commodity over a sufficient period of time to avoid the local drops in prices often occurring when large quantities of a commodity were thrown on the market at harvest time." This had particular meaning for the dairy industry since production in the spring during what is called the "flush season" often outstrips demand, causing depressed milk prices. Prior to the legislation in the early 1930s, cooperatives had difficulty negotiating higher prices for fluid milk with dealers during this period of the year.

According to the Nourse report (Federal Milk Order Study Committee 1962), the AAA of 1935 and the AMAA of 1937 expressed four broad purposes:

1. To bring all distributors (handlers) in a prescribed marketing area under the scope of the regulatory mechanism (the order).
2. To place all handlers in the same competitive position by requiring the use of minimum prices for milk entering the same use (classified pricing).
3. To provide for a uniform price for all producers (marketwide pooling).
4. To extend classified pricing and pooling plans to all handlers and producers in a prescribed marketing area in order to overcome instability in fluid milk pricing.

The Nourse report further commented that "marketing orders provided the means of extending uniform opportunities and responsibilities to (and enforcing them upon) the entire market, rather than certain handlers only."

Despite concerns for the public good, however, the overriding objective of Congress in this early legislation was to improve the economic conditions of dairy farmers.

Key Provisions of an Order

As stated earlier, FMMOs represent an agreement between the secretary of agriculture, producers, and handlers of Grade A milk in a defined geographic region. Under federal order language, a handler is any person or cooperative association of dairy producers that operates one or more dairy processing plants and markets fluid milk. A handler that only processes Class III products (e.g., cheese) and not fluid milk cannot be regulated under FMMOs. The geographic region represents a marketing area where consumers purchase dairy products. This region typically centers around urban locations and normally contains bottling plants. The boundary line for the marketing area is usually drawn where there are few route sales that cross into other marketing areas. Another characteristic of the marketing area is the producers that regularly supply Grade A milk to that market. These producers are identified with the market and therefore share in the benefits of Class I fluid milk sales, as well as the costs of marketing lower-valued surplus milk. These producers may or may not physically reside in the area governed by the order. For example, producers in Wisconsin are regulated in several federal orders since milk is regularly shipped out of state. Remember, marketing areas are not defined according to where milk is produced, but rather where it is marketed. Also, producers are only affected by an FMMO if they sell their milk to a handler that is regulated.

Federal orders are voluntary agreements that are usually initiated by dairy farmers (via their cooperatives) since the purpose of a milk order is to provide dairy farmers with a marketing plan under government supervision. Producers must approve a new order or an amended order by vote in a referendum. The AMAA of 1937 allows cooperatives to vote in referendums on behalf of their members through bloc voting. Bloc voting is the vote of a bona fide cooperative association and must be accepted by the secretary of agriculture as the vote of all members of the cooperative who are eligible to vote on a prospective order. Bloc voting has been criticized in recent years since some members feel that the vote of cooperatives may not fully represent all members.

Each order is administered by a market administrator (MA), who is an agent of the secretary of agriculture. According to the USDA, the MA's main duty is to ensure that handlers properly account for how they use milk and pay producers according to provisions of the order (USDA, AMS 1989). The MA has a staff that investigates and audits the books of handlers in order to ensure that the required payments are made to producers. Handlers are required to make monthly reports to the MA. The MA in turn is required to report to the public each month classified prices, how milk was used in the order, and the blend price.

Producers, their milk, and plants that process that milk must be identified regularly with a particular order. Performance standards for these players are defined in each order to ensure that the benefits to be derived from the order (Class I fluid sales) are shared equitably. In other words, handlers that regularly supply a particular market are regulated in order to ensure that the producers they buy milk from receive the benefits of FMMOs. For example, producers or co-ops that ship milk to, say, the Southwest Plains order can't have their milk pooled on that market if they don't regularly supply that market (more about this later). In addition, plants must be identified with a particular order. These plants are subject to the terms and provisions of a particular order and are therefore referred to as "fully regulated." It should be stressed that it is the handlers that are regulated under FMMOs, not the producers.

Federal orders have performance standards for three types of plants.

The first is a distributing pool plant, which has a significant portion of its milk receipts sold for fluid uses. This is typically a bottling plant. The second type of plant is a supply plant, which is typically a manufacturing plant. To qualify to be fully regulated on the order, supply plants must transfer or divert some of their milk receipts to a distributing pool plant. And finally, the third type of plant is a cooperative association plant that typically processes surplus milk. Since it is owned by farmers, it doesn't have the same performance standards that a supply plant has in order to qualify on an order.

The major provisions of federal orders are classified pricing, whereby milk is priced according to its end use, and marketwide pooling, whereby receipts from handlers are paid into a pool and an average price is paid back to producers. These concepts are discussed next.

Classified Pricing

Classified pricing essentially establishes (1) one price for milk entering fluid uses and (2) prices for the balance of milk that is used for manufacturing purposes. The Nourse report cites the overriding objectives of classified pricing as ensuring an adequate supply of milk for fluid use and establishing orderly marketing conditions. Classified prices represent minimum prices that handlers must pay for milk based on its end use. Handlers often pay above these minimum prices in some orders and at certain times of the year depending on local market conditions.

Classified pricing creates a higher value for milk used for bottling purposes. The Class I price for milk refers to milk used for bottling purposes. The higher price for bottled milk has been justified on two grounds (USDA, AMS 1989). First, milk to be used for fluid uses must be produced under costly sanitary conditions. Higher prices for Grade A milk are used to encourage producers to switch from producing manufacturing grade milk—otherwise known as Grade B milk, which has a lower cost of production. Second, milk in fluid form is more costly to transport (due to the high concentration of water in milk) than manufactured dairy products. Therefore, higher prices are required in order to pay for milk to be hauled long distances to processing plants in urban areas. A third justification could be that consumers are less sensitive to changes in the price of fluid milk than they are to changes in cheese prices. Therefore, higher prices for bottled milk could be set without a significant drop in consumption (Chapter 2).

Classified pricing also helps in the disposition of surplus supplies of fluid-eligible milk, thereby stabilizing fluid milk prices. Fluid sales generally peak in the fall months, whereas production of milk peaks in the spring and declines through the summer months. Therefore, in order to have an adequate supply of fluid milk all year long, Grade A milk production must be produced in surplus of what is required in the spring months in order to have adequate supplies in the fall. (Remember, Grade B milk cannot be used for bottling purposes.) In addition, bottling is not constant from one day to the next; it usually peaks at the beginning and end of the week (Nicholson 1994). Thus, in order to have an adequate supply of fluid milk throughout the week, there must be an excess of Grade A milk purchased. Classified pricing helps in balancing the market by creating a lower value for surplus milk in order to enhance its disposition.

The federal order program began with just two class uses: Class I for fluid uses and Class II for all other surplus milk. In the 1940s, some orders, such as the New York order, had several class prices. But by August 1974, most orders contained just three classes for milk. The Class II price was for milk used for soft manufactured dairy products such as yogurt, ice cream, cottage cheese, and cream products. The Class III price was for milk used in the processing of hard manufactured dairy products such as cheese, nonfat dry milk, and butter. It was the lowest classified price as it represented a price for surplus milk.

A fourth classification of milk began to be used on November 1993 in three Northeast orders. Dairy cooperatives who owned nonfat dry milk processing facilities began to complain to the USDA that they could not continue to afford to pay the same price for milk as cheese processors. The problem was that the price of cheese strengthened relative to the price of nonfat dry milk. Therefore, milk used in the manufacture of cheese had a higher market value than milk going into the production of nonfat dry milk. In addition, processors in the Northwest had to compete with California processors of nonfat dry milk who paid a much lower price for milk used in the production of nonfat dry milk. The result was the approval of a Class IIIa price for milk used in the manufacture of nonfat dry milk. The Class IIIa price is now the lowest classified price in all federal orders.

Manufacturing Grade Price of Milk

As stated earlier, Grade A milk must be produced in excess of fluid needs in order to ensure that an adequate supply of fluid-eligible milk is available all year round. The problem then is how to price this surplus milk. Remember, surplus milk is a federal order concept defined as Grade A milk in excess of bottling needs in a local market. Grade A surplus milk cannot be priced too high since the market must be willing to dispose of all this milk in the form of manufactured dairy products. On the other hand, it cannot be priced too low since that would adversely affect producer income.

Historically, Class III milk has been valued in federal orders by three methods. The first method used product price formulas whereby the wholesale value of surplus milk was calculated using dairy product prices, product yield factors, an assumption of plant profits, and a make allowance. The make allowance approximates the cost of manufacturing various dairy products from raw milk. The second method used a survey of handlers that manufactured dairy products in markets not regulated by federal orders (called "nonregulated handlers"). This survey was used to determine a competitive pay price. The market for Grade B milk is one such market since handlers that purchase Grade B milk are not regulated under federal orders. A third method used a combination of the above. For example, the Chicago order prior to the 1960s valued manufacturing grade milk based on (1) 18 condenseries' price and (2) a butter–nonfat dry milk formula (USDA 1954).

The advantage of a product formula is that it can be derived without a formal survey. The problem, however, is that plant profits and the cost of manufacturing dairy products are not uniform among plants and change over time. The advantage of pricing Grade B milk in nonregulated markets is that such a price is based on competitive market supply and demand conditions. The problem with this pricing method, however, is that the market for Grade B milk has been shrinking over time.

By the early 1960s, Midwest manufacturing plants began to question the wisdom of having several different methods of pricing surplus milk under federal orders. They complained that such prices were much lower in the Northeast than in the Midwest. Therefore, manufactured dairy products from Midwest plants were less competitive. Their argument had merit since manufactured dairy products by that time were trading on a national market. What was needed was a uniform price for surplus milk that was applicable to all federal orders. Such a price would have to be derived from competitive pricing in an unregulated market in order to avoid the inherent problems of using product price formulas that had been prevalent up until that time.

The solution was the Minnesota-Wisconsin (M-W) price series for manufacturing grade milk, which was initially adopted in the Chicago order in 1961. This series is based on competitive market conditions in the unregulated market for Grade B milk in Minnesota and Wisconsin, where about half of the nation's supply of Grade B milk is produced. The M-W reflected a competitive market price for surplus milk and served as an explicit link between Grade B and surplus Grade A milk prices. This link is due to the fact that manufactured dairy products compete in a national market and can be produced from either Grade B milk or surplus Grade A milk.

The M-W was computed by the National Agricultural Statistics Service on or before the fifth of each month and represents manufacturing grade milk delivered f.o.b. the plant or receiving station, whichever is the customary point for determining the milk price to producers (USDA, NASS 1992). It is estimated before hauling costs, producer assessments, and checkoffs are deducted, and includes quantity, quality, protein, and other premiums paid to producers. The data used to estimate the M-W are for Grade B milk only and do not reflect Grade A milk diverted to manufacturing uses.

The M-W was replaced by the basic formula price (BFP) on June 1, 1995, as the price for manufacturing grade milk under all federal orders. The USDA replaced the M-W because it was concerned that the M-W was no longer statistically reliable since Grade B marketings had declined from 33 percent of total U.S. marketings in 1960 to just 8 percent in 1990.

The BFP is computed each month in two parts. The first part consists of computing a base month price. This price is determined using reports from about 100 plants in Wisconsin and 70 in Minnesota. These plants purchase about 80 percent of all Grade B milk sold in these two states. About 95 percent of the dairy products produced by these plants are cheese. They report the total pounds of Grade B milk received from producers, total pounds of milkfat in the milk, and total dollars paid to producers. This survey provides an estimate of the price of manufacturing grade milk for a full month.

The problem with this statistical estimate is that it takes over a full month to construct it. Results from a July survey, for example, won't be available till the end of August. Therefore, in order to provide more timely information, a butter/powder/cheese product formula is used in the BFP to update or adjust the base month survey price. Through the formula, the new BFP recognizes changes in the value of milk used to manufacture cheddar cheese, butter, and nonfat dry milk from the survey month (preceding month) to the current month. These changes are computed and then weighted by production (the cheese price has a greater weight than the butter/nonfat dry milk price due to greater production in the survey states). The change in commodity prices is then added to the base month price calculated for the preceding month.

To put things in perspective, let's assume the September base month price is $12.19 and we want to estimate the BFP for October (to be released November 5). The BFP is an estimate of the manufacturing grade price of milk since the true price for October manufacturing grade milk (the survey price used in the base month price) won't be released till December 5. To calculate the BFP we need to first calculate the basic formula price adjustment by estimating the difference between dairy commodity prices in September and October. These changes are then weighted by relative production of the commodities. Let's assume that for our example this weighted average change is 42 cents. The BFP for October then is $12.61 ($12.19 + $0.42).

Fluid-Grade Price of Milk

Grade A milk used for bottling purposes is referred to as "Class I milk" in federal order language. It is the highest-value use under the classified pricing system and represents a minimum price that milk handlers must pay for milk used for bottling purposes. In the early years of federal orders, two Class I price formulas were used: (1) a differential over the price of milk for manufacture and (2) a base price adjusted by indexes of various economic factors. The latter was used mainly in the Northeast and was based on a number of factors including department store sales, sales of nondurable goods, and farm wage rates (USDA 1954). In addition, seasonal factors and a supply-demand adjustment were used to adjust the Class I prices in federal orders to market conditions.

Class I milk prices have historically been set by the secretary of agriculture in individual orders following a public hearing. The objective in setting this price was to ensure an adequate supply of fluid milk to the local market. The Class I price is set above the Class III price by a Class I differential, which varies from order to order. So why is the Class I price purposely set above the Class III price? The logic used in the past was reiterated by the secretary of agriculture in a recent amplified final decision as follows, "because some milk is produced just about everywhere, the Class I differential in any particular marketing area merely has to be high enough to bring forth adequate supplies of locally produced milk together with supplemental supplies from other areas" (USDA, AMS 1994b, p. 6).

Thus if you are in a fluid-deficit market, the Class I differential should be higher than in other markets that regularly produce surplus Grade A milk. The higher price would either encourage more local production or help pay the cost of transporting milk into that market.

When federal orders were first implemented, Class I prices were based on local conditions since fluid milk production and processing were entirely local in nature. However, as the federal highway system developed and fluid milk production and processing became more regionalized, changes in the determination of Class I prices were needed. By the early 1960s, Class I differentials in individual orders were set in relation to Class I differentials in the Chicago order, which in turn set its Class I differential in relation to the distance of that market to the surplus region of Eau Claire, Wisconsin. At the time, the Upper Midwest had a substantial surplus of Grade A milk in excess of fluid needs, whereas most southern states were net importers of fluid milk.

What then developed out of a number of federal order hearings in the early 1960s was a formula whereby most orders east of the Rocky Mountains set their Class I differentials according to a distance differential equal to 15 cents per hundredweight for each 100 miles the base zone city in the order was from Eau Claire, Wisconsin. These distance differentials were intended to represent the cost of transporting milk from surplus areas to deficit areas to avoid milk shortages (U.S. General Accounting Office 1988).

This formula for setting Class I differentials, however, was replaced by new Class I differentials mandated by the Food Security Act of 1985.

Since some orders had an increase in their differential and others did not, there is no longer a defined relationship between Class I differentials and the distance an order is from Eau Claire (Table 6.1). In general, however, these differentials still increase relative to the distance an order is from Eau Claire.

To see how Class I prices are set, let's look at a few real-world examples. Handlers using milk for fluid use in the Southwest Plains order have to pay at least the Class I price, which is equal to the BFP two months prior plus a Class I differential of $2.77 per hundredweight. Oklahoma City, the central base point for the Southwest Plains order is exactly 838 miles to Eau Claire, Wisconsin. However, if you are a handler in Miami, Florida, you are required to pay at least the Class I price, which is equal to the BFP two months prior, plus a Class I differential of $4.18 per hundredweight. That is because Miami represents the base point for the Southeast Florida order, which is 1,618 miles from Eau Claire.

Class I differentials don't change seasonally; they are the same each month. Also, the lagged BFP used in the formula for the Class I price was designed to provide fluid bottlers 25 days advance notice of what the minimum fluid milk price will be. The logic for this advance notice is to give bottlers time to bid for contracts and to announce price changes to their wholesale distributors.

Cooperatives receive special treatment from authority granted under the Capper-Volstead Act. They can combine their market strength by forming a federation of cooperatives also known as a "marketing agency in common" and bargain with fluid processors and other handlers for prices above minimum classified prices. The proceeds of these higher prices are pooled in a superpool, and the proceeds distributed after removing marketing costs. For example, the difference between these higher market prices and the minimum classified prices are referred to as "overorder premiums." It is important to note that they are collected and distributed outside of the federal order system. For example, the Central Milk Producers (CMP) Cooperative bargains for higher milk prices on behalf of cooperatives in the Midwest. The Class I overorder premium negotiated by CMP in the Chicago market on August 1994 was $1.92 per hundredweight (USDA, AMS 1994c).

Table 6.1. Change in Class I differentials before and after the Food Security Act of 1985

Federal Marketing Order	After May 1, 1986	Before May 1, 1986	Increase
		($/cwt)	
Middle Atlantic	3.03	2.78	.25
New England	3.24	3.00	.24
New York-New Jersey	3.14	2.84	.30
Upper Florida	3.58	2.85	.73
Georgia	3.08	2.30	.78
Tennessee Valley	2.77	2.10	.67
Tampa Bay	3.88	2.95	.93
Southeastern Florida	4.18	3.15	1.03
Chicago Regional	1.40	1.26	.14
Southern Illinois	1.92	1.53	.39
Ohio Valley	2.04	1.70	.34
Eastern Ohio-Western Pennsylvania	1.95	1.85	.10
Southern Michigan	1.75	1.60	.15
Michigan Upper Peninsula	1.35	1.35	0
Louisville-Lexington-Evansville	2.11	1.70	.41
Indiana	2.00	1.53	.47
Central Illinois	1.61	1.39	.22
Greater Kansas City	1.92	1.74	.18
Nebraska-Western Iowa	1.75	1.60	.15
Upper Midwest	1.20	1.12	.08
Black Hills, South Dakota	2.05	1.95	.10
Eastern South Dakota	1.50	1.40	.10
Iowa	1.55	1.40	.15
Alabama-West Florida	3.08	2.30	.78
New Orleans-Mississippi	3.85	2.85	1.00
Greater Louisiana	3.28	2.47	.81
Memphis, Tennessee	2.77	1.94	.83
Nashville, Tennessee	2.52	1.85	.67
Paducah, Kentucky	2.39	1.70	.69
Southwest Plains	2.77	1.98	.79
Central Arkansas	2.77	1.94	.83
Lubbock-Plainview, Texas	2.49	2.42	.07
Pacific Northwest	1.90	1.90	0
Texas	3.28	2.32	.96
Central Arizona	2.52	2.52	0
Texas Panhandle	2.49	2.25	.24
Western Colorado	2.00	2.00	0
Southwestern Idaho-Eastern Oregon	1.50	1.50	0
Great Basin	1.90	1.90	0
Eastern Colorado	2.73	2.30	.43
Rio Grande Valley	2.35	2.35	0

Source: Federal Market Administrator, Tulsa, Oklahoma.

Pricing Soft Manufactured Dairy Products

Grade A milk used in the processing of soft manufactured dairy products, such as cottage cheese, yogurt, creams, ice cream mixes, fluid cream products, and other semisolid dairy products, is referred to in federal order language as "Class II milk." Class II prices are set above Class III prices in order to attract a sufficient quantity of milk for Class II needs away from surplus milk uses. The Class II price in the past was determined by a simple formula equal to the Class III price lagged two months plus a Class II differential, which was set at 10 cents per hundredweight. The Class II differential gave way to a more complicated formula based on the gross value of milk used in the manufacture of cheddar cheese, butter, and nonfat dry milk. The prices of these manufactured dairy products and relevant yield factors were used to approximate the gross value of milk (USDA, AMS 1993). That formula, however, was replaced by a new formula in 1995 equal to the BFP lagged two months plus a Class II differential equal to 30 cents per hundredweight. Class II prices provide the industry with a 25-day advance notice announced at the same time as Class I prices. The previous formula was considered too complex to forecast each month by Class II processors and only reflected a 15-day advance notice.

Location Adjustments

As already discussed, formulas for Class I differentials were originally set to reflect the cost of transporting milk from the surplus regions in the Upper Midwest to other parts of the country. They were also to reflect a price that would ensure a local supply of milk, whether produced locally or imported from other regions of the United States. However, within any particular order, the Nourse report noted that "providing substantially equal raw product costs to all competing handlers ... requires that different prices for Class I milk be established for various locations within any milkshed."

In general, in all federal orders there is a major population center where most of the fluid bottling plants are located. In addition, dairy producers and manufacturing facilities are located on the outer fringes of the order away from the population centers. Administered prices under federal orders would therefore require that some mechanism be set up to allow for the transportation of milk from the production/surplus region of an order to the population center where the fluid plants are located. Such a system must also ensure that handlers in one order are on an equal footing with handlers in other orders.

Zone price adjustments were therefore established in federal orders to account for the costs of transporting milk from the surplus production area of an order into the population centers. As an example, Oklahoma City is the base point of the Southwest Plains marketing order and is part of zone 1. Springfield, Missouri, also part of the same order, is considered a surplus region and lies 286 miles northeast of Oklahoma City in zone 7. Under the rules of the Southwest Plains order, the Class I price and the uniform blend price in zone 7 are 58 cents per hundredweight less than in zone 1. Thus, dairy producers in Springfield essentially pay the cost of transporting surplus fluid milk to Oklahoma City. It is a basic tenet of all federal orders that farmers pay the cost of transporting milk from the farm to the processing plant.

Zone charges are also used to align Class I prices within and across various orders. For example, in the Chicago order, zone 12 refers to Eau Claire and represents the cost of transporting milk between Eau Claire and Chicago.

Marketwide Pooling

The concept of marketwide pooling has been used in federal orders since their creation. All handlers in an order must pay at least classified prices for milk into a marketwide pool. The MA then blends the proceeds of the pool according to use and pays out a uniform price to all producers who marketed their milk in the order. This price is referred to as "uniform price" since all producers receive the same price (before hauling costs and zone charges are accounted for) no matter which handler their milk was shipped to. So, even if farmer A's milk went to a cheese plant and farmer B's milk went to a fluid plant, both farmers would receive the same price. This is a critical concept since it ensures that all producers equitably share in the benefits of the higher-valued fluid markets.

Pool Example

A simple demonstration of how a marketwide pool works is presented in Table 6.2 using a fictitious pool for a fictitious month. Let's assume that there are three handlers in this pool: Hiland, a fluid bottler, Mid-Am, a dairy cooperative, and Kraft, a cheese manufacturer. In this example, the Hiland plant is considered a distribution plant because it meets certain performance standards stipulated in the order. These performance standards vary from order to order. In our example Hiland qualifies to market milk in this fictitious order by having a significant proportion of its producer milk receipts (96 percent) processed and sold for Class I uses. The remainder of Hiland's milk was processed into Class II and Class III uses. We will assume that Hiland buys all of its milk from Mid-Am members. The Kraft plant is considered a supply plant. Producers that supply milk to this plant "qualify" to market milk in this order (and thus benefit from Class I sales) by shipping part of their milk to a fluid plant. Therefore, 50 percent of Kraft receipts from its producers was diverted to Hiland as Class I sales in order to qualify on the order. The rest of Kraft's producer receipts were manufactured into cheese.

Table 6.2. Computation of an example marketwide pool

	Hiland		Mid-Am		Kraft		Total market	
Handler	(cwt)	(%)	(cwt)	(%)	(cwt)	(%)	(cwt)	(%)
Producer receipts								
Class I	38,400	96	0	0	25,000	50	63,400	36
Class II	400	1	5,500	6	0	0	5,900	3
Class III	1,200	3	80,000	90	25,000	50	106,200	59
Class IIIa	0	0	3,500	4	0	0	3,500	2
Total	40,000	100	89,000	100	50,000	100	179,000	100

	Hiland		Mid-Am		Kraft		Total market	
Handler value	$	$/cwt	$	$/cwt	$	$/cwt	$	$/cwt
Class I	556,800		0		362,500		919,300	
Class II	5,040		69,300		0		74,340	
Class III	14,400		960,000		300,000		1,274,400	
Class IIIa	0		36,750		0		36,750	
Total	576,240		1,066,050		662,500		2,304,790	
Average classified value		14.41		11.98		13.25		12.88

Note: Classified prices ($/cwt) used in this analysis are as follows: Class I: $14.50; Class II: $12.60; Class III: $12.00; and Class IIIa: $10.50.

Finally, the Mid-Am plant is considered a cooperative association plant. Since this plant is owned and operated by farmers and serves a balancing function in the market, it does not have the same performance requirements as the Kraft plant in order to qualify on the order. In our fictitious example, Mid-Am must market part of its producer receipts to a distribution plant. Mid-Am qualifies since it supplies all of Hiland's raw milk needs (38,400 cwt). All three handlers in this example order are considered fully regulated in that they meet performance guidelines stipulated by the federal order and are subject to the terms and provisions of the order.

All three handlers in this pool report to the MA the pounds of milk received from producers and how this milk was used during the month. The MA then computes the uniform blend price based on previously announced class prices and information reported by handlers. In this example, a blend price of $12.88 per hundredweight was computed by multiplying the class price by the pounds of milk used for each handler and dividing the sum by the total pounds of milk received from producers in the order.

Producer Settlement Fund Example

In theory, handlers pay the classified value of milk (the class price times the amount of milk used for that class purpose) into a marketwide pool based on their own utilization and then draw from the pool the blend price that they pay producers. In reality, however, handlers only account to the pool for the net transaction. For example, the Hiland plant received 40,000 hundredweight of milk of which 96 percent was used for Class I purposes, 1 percent for Class II purposes, and 3 percent for Class III purposes. Given classified prices of $14.50, $12.60, and $12 for Class I, II, and III, the classified value of this milk was $576,240 (Table 6.3). However, Hiland is only directly accountable to producers for $515,037 (40,000 · $12.8759). The difference between the classified value of milk and the amount paid producers is $61,203, which was paid into a producer settlement fund. The classified value of Mid-Am receipts was $1,066,050, but Mid-Am was only accountable to producers for $1,145,957. The classified value was lower than the amount owed producers since most of Mid-Am's milk is used for the lower-valued Class III use. Mid-Am will then draw $79,907 from the producer settlement fund. Kraft will also pay $18,704 into the producer settlement fund since part of its milk is bottled at the Hiland plant and is valued at the higher Class I price. All three transactions in the producer settlement fund cancel each other out by the end of the month. In general, handlers with higher-valued uses for milk (Class I and II) pay into the producer settlement fund, whereas handlers with lower-valued uses (Class III and IIIa) draw from the fund.

Allocation of Milk

The MA calculates the uniform blend price each month based on receipts of milk from producers regulated in the MA's order. If milk is shipped into the order from producers regulated on another order, or from nonregulated handlers, those milk receipts are removed from the pool before the uniform price is calculated. This process of removing nonproducer milk receipts and assigning those receipts to alternative class uses for each handler is called the "allocation process." The MA accomplishes the allocation procedure for each handler every month based on the utilization report provided by each handler in the order.

The first step in the allocation procedure is to account for all receipts of milk and milk products in packaged, bulk, fluid, and nonfluid form. These would include dairy ingredients such as nonfat dry milk used in the production of ice cream. The handler then reports how these products were used and the beginning and ending inventories of dairy ingredients and products. With this information, the MA removes all nonproducer milk (i.e., milk from other orders, milk from a nonregulated handler, or the milk equivalent of dairy ingredients) from the pool and allocates the producer milk to classified uses.

Table 6.3. Computation of an example producer settlement fund

	Blend price paid producers ($/cwt)	Amount paid producers ($)	Classified value of milk ($)	Producer settlement fund ($)
Hiland	12.88	515,037	576,240	61,203
Mid-Am	12.88	1,145,957	1,066,050	(79,907)
Kraft	12.88	643,796	662,500	18,704
Total market	12.88	2,304,790	2,304,790	0

The purpose of this monthly exercise by the MA is to preserve the benefits of Class I sales for those producers who regularly supply the market each month.

MA Deductions and Charges

Each month the MA may hold back funds from the producer settlement fund and charge handlers fees in order to carry out the mission of the federal order. The direct cost of federal orders is financed by the industry and does not use taxpayer funds.

A reserve fund is created each month by the MA in order to meet possible adjustments in handler costs based on forthcoming audits. The MA pools producer receipts and estimates the uniform blend price each month based on data provided by handlers. This data, however, is unaudited at the time of the calculation. The MA's office may subsequently audit the handler records submitted and find mistakes that may result in a draw from or payment into the producer settlement fund. In order to facilitate these transactions after the calculation of the uniform blend price, a reserve fund is created each month by withdrawing 4–5 cents per hundredweight from the producer settlement fund.

An example of a uniform price calculation for the Southwest Plains order is provided in Table 6.4. Note that 4.252 cents per hundredweight

Table 6.4. Computation of the July 1994 uniform blend price for the Southwest Plains marketing area

	Utilization (%)	Receipts (cwt)	Price ($/cwt)	Value ($)
Class I	37.03	1,226,392	14.28	17,512,878
Class II	16.53	547,406	10.35	5,665,652
Class III	40.01	1,325,177	11.41	15,120,274
Class IIIa	6.43	213,099	10.13	2,158,689
Weighted average price			12.21	40,474,189
Add location adjustment				867,858
Add ½ ending fund reserve				166,587
Less new reserve			.04252	140,830
Uniform blend price at 3.5% butterfat			12.49	41,367,804

Source: Market Administrator, Southwest Plains Marketing Order.

was withdrawn from the pool in order to fund a reserve ($140,830). Also note that at the time of the calculation of the uniform price (the last day of July 1994), one-half of the unused portion of the previous month's reserve fund ($166,587) was transferred into the producer settlement fund and was used in the blend price calculation.

The MA charges handlers regulated in the order a 4-cents-per-hundredweight fee that goes into an administrative fund. These funds are used by the MA to pay for such things as personnel, office space, and office supplies. In addition, nonmember producers who are not represented by a regulated handler or a cooperative must pay a 6-cents-per-hundredweight fee that goes into a market service fund. This fee is higher than for regulated handlers since milk from nonmember producers must be tested by the MA and their bulk tanks calibrated, and the producers are eligible to receive monthly market reports. Producers that market their milk through a cooperative, for example, are only charged the administrative fee since their cooperative tests their milk for components and quality, and the management of the cooperative receives all market information on behalf of the members.

Marketing Milk between Orders

A cornerstone of an individual marketing order is that producers who regularly supply the Class I needs of that market are entitled to the proceeds from the higher-valued Class I sales. That is the purpose of the allocation procedure discussed earlier. But what happens when an outside producer markets milk into an order so as to provide that pool with sufficient milk to meet consumer needs? Why isn't the producer entitled to a part of the Class I milk sales in that order? If the producer is not, wouldn't there be an incentive for an unregulated handler (who is not required to pay minimum prices to producers) to regularly move milk into an order with a relatively high Class I use and thereby benefit from the Class I sales?

Compensatory Payments

When orders were first created, milk rarely moved between orders. But that changed with the use of farm bulk tanks, insulated tanker trucks and the creation of the interstate highway system. To illustrate the complexities of how milk is moved between federal orders, let's review the *Lehigh Valley* decision of the U.S. Supreme Court (USDA 1964). The decision involved the Lehigh Valley Cooperative of Pennsylvania, which was charged a compensatory payment after marketing milk into the New York–New Jersey marketing order. A compensatory payment was defined as the difference between the minimum fluid milk price and the surplus price (Class III). It was used at the time to compensate producers who regularly supplied this market. The New York–New Jersey order considered Lehigh Valley Cooperative a nonpool handler and its milk "outside" or "other source milk." This case challenged the legality of compensatory payments.

In order to illustrate how compensatory payments work, let's use the example presented by Justice Harlan, who delivered the opinion of the U.S. Supreme Court. While the prices used in this example were modified to reflect current prices, the example is otherwise the same. Suppose we have an order with only Class I and Class III uses. Let's further assume class prices of $14 per hundredweight for Class I uses and $10 per hundredweight for Class III uses. The calculation of a uniform blend price of $12 per hundredweight is presented in Table 6.5. This example assumes that the pool plants in this order use 2,000 hundredweight of milk for fluid purposes and 2,000 hundredweight of milk for manufacturing purposes and that this milk is regularly supplied by area producers.

Now, let's assume that 500 hundredweight of milk is brought in from outside the order, is used for Class I purposes, and is supplied by an unregulated handler. The uniform blend price is calculated based entirely on milk from producers who regularly supply the market and are there-

Table 6.5. Example computation of the uniform blend price from the *Lehigh Valley* decision

	Classified price ($/cwt)	Use (cwt)	Value ($)
Class I	14.00	2,000	28,000
Class III	10.00	2,000	20,000
Total pool milk		4,000	48,000
Uniform blend price	12.00		

fore regulated on this market. Let's suppose the outside milk displaces Class I sales but leaves total producer receipts for the pool unchanged. This is illustrated in Table 6.6, where Class I use is reduced by the 500 hundredweight to reflect the Class I displacement from the outside milk, and Class III use is increased by 500 hundredweight. Total marketings by producers in the pool remain the same at 4,000 hundredweight. In this example the outside milk effectively reduced the pool value and lowered the uniform blend price by 50 cents per hundredweight.

Now let's use a compensatory payment to compensate pool producers for the loss associated with outside milk sales. This is illustrated in Table 6.7, where a compensatory payment of $4 per hundredweight (equal to the difference between the Class I and III price) is paid to pool members on the volume of outside milk. This offsets the pool loss and thereby maintains a $12-per-hundredweight blend price.

The Lehigh Valley Cooperative prevailed in the case under the

Table 6.6. Effect of 500 hundredweight of Nonpool Milk for Class I Use on the Uniform Blend Price

	Classified price ($/cwt)	Use (cwt)	Value ($)
Class I	14.00	1,500	21,000
Class III	10.00	2,500	25,000
Total pool milk		4,000	46,000
Uniform blend price	11.50		

Table 6.7. Effect of a compensatory payment on 500 hundredweight of nonpool milk on the uniform blend price

	Classified price ($/cwt)	Use (cwt)	Value ($)
Class I	14.00	1,500	21,000
Class III	10.00	2,500	25,000
Compensatory payment (nonpool milk)	4.00	500	2,000
Total pool milk		4,000	48,000
Uniform blend price	12.00		

premise that use of compensatory payments imposed "insuperable trade restrictions on the entry of nonpool milk into a marketing area." As a result of this case, all federal orders were amended in a final decision issued by the secretary of agriculture in 1964. This decision suspended the financial obligation of nonpool milk handlers or producers to pay compensatory payments.

The challenge to the secretary of agriculture in revising all FMMOs in order to make them in compliance with the *Lehigh Valley* decision was how to continue to protect pool producers, those that regularly supplied Class I needs to a pool, while avoiding trade restrictions. The method adopted in the secretary's final decision is discussed later in an example.

Understanding how milk is shipped between orders since the *Lehigh Valley* decision is complicated. A major objective of FMMOs is to keep the Class I market for locally produced milk. The major change since then is that unregulated handlers cannot be assessed a compensatory payment if they decide to ship milk to a handler in a regulated market. Two examples are provided here in order to better illustrate current regulations. Let's start with shipments between regulated handlers.

Milk Shipped from Regulated Handlers

The federal orders have been designed and modified by court challenges to protect local producers that regularly supply a Class I market. They do that by protecting Class I prices and usage or, at least, by not allowing the local pool to be diluted.

Let's suppose that Borden of Little Rock decided to ship 2,000 hundredweight of fluid milk to the Hiland dairy plant in Springfield, Missouri. This Hiland plant is regulated on the Southwest Plains order. Since Borden is regulated on the Arkansas order, it is considered an "other order plant" by the Southwest Plains order. In addition, the milk shipped to Hiland is pooled on the Arkansas order and is not included in the calculation of the blend price in the Southwest Plains order. However, the questions at hand are, how are the producers on the Southwest Plains order protected and how is this volume of milk pooled on the Arkansas order?

Two approaches are available since the *Lehigh Valley* decision. First, both handlers (Hiland and Borden) can decide that the shipment was for Class III purposes. In that case, the Borden shipment of 2,000 hundredweight would be pooled on the Arkansas order and allocated to Class III purposes. Southwest Plains producers would benefit by keeping their Class I milk sales in their pool. Another approach would be to use a pro rata approach, whereby the Borden plant would benefit from the higher Class I sales in the Southwest Plains order. Producers on the Southwest Plains order would not be affected since their percent utilization would be maintained. To see how this would be accomplished, let's assume that the Southwest Plains order had an 80 percent Class I use and a 20 percent Class III use after removing the Borden shipment from the Southwest Plains pool. Under the pro rata approach, Borden could benefit from the sale since the MA in the Arkansas order would credit Borden with 1,600 hundredweight of Class I sales (2,000 hundredweight times 80 percent) and only 400 hundredweight of Class III sales (2,000 hundredweight times 20 percent).

Milk Shipped from Unregulated Handlers

Now what would happen if Hiland buys milk from an unregulated handler for Class I use? Prior to the *Lehigh Valley* decision, the unregulated handler had to make a compensatory payment into the pool, and that milk would be stripped from the pool in the allocation process. Today, there are two options available to handlers. First, both the unregulated and regulated handler could agree that the milk shipped would be used for Class III purposes. A price for this milk would be agreed upon between the two handlers, a price that is not enforced by the MA. During the allocation process the MA would strip this milk away from Class III uses since only producer milk can be used to calculate the uniform blend price. The milk in the order from producer receipts would then be "up allocated" to higher uses, thus raising the uniform blend price for local producers after nonproducer milk was removed.

Under the second option, both handlers would agree that the milk would be used for Class I purposes and would request that the milk be handled pro rata. Both handlers would still reach an agreed-upon price that could not be enforced by the MA. In addition, this milk would be effectively removed from the order during the allocation process in order to calculate the uniform blend price. The major change from the first option is that the *regulated handler* would be accountable to the pool for a compensatory payment, not the out-of-order unregulated handler. The amount of the compensatory payment would depend on whether or not the regulated handler was short of milk for fluid needs. In any case, local producers would not be affected since the pool value would be preserved.

If any nonregulated milk is assigned to Class I use, then the regulated handler (the handler on the order that buys the nonregulated milk) is responsible for paying a compensatory payment into the pool equal to the difference between the Class I price and the blend or Class III price (depending on the order) on the volume of milk assigned to Class I uses. Thus it is the regulated handler that is responsible to the pool if nonproducer milk is purchased and assigned to Class I uses. The MA has no authority to assess the out-of-order nonregulated handler for a compensatory payment.

This process of down allocating nonproducer milk receipts (assigning this milk to lower-valued class uses) and assigning a compensatory payment to regulated handlers effectively discourages importing unregulated milk for fluid uses into an order (McDowell et al. 1988). This is especially true if sufficient quantities of producer milk are available in relation to Class I needs on the order.

Creating or Amending an Order

FMMOs were initially created with the capacity to change over time as market conditions change. They are therefore constantly revised in order to keep them current with changes in market conditions. These changes occur administratively through public hearings, court action, or changes in federal law.

The AAA of 1937, as amended, authorized FMMOs and defines the role of the federal government in administering them. The USDA is responsible for administering the federal order program and in judging the merits of proposals made at public hearings to create or change them. It is through the hearing process that orders are created and amended to reflect changing market conditions.

A new marketing order is normally proposed by dairy farmers through the actions of their cooperatives. That is not a requirement, however, as any individual or even the secretary of agriculture can request a new marketing order. A written request for a new order is forwarded to the dairy division of the Agricultural Marketing Service (AMS) of the

USDA. The USDA then makes a preliminary investigation into the merits of the proposal. If it has merit, additional proposals are solicited from the dairy industry and the public, and a prehearing study is conducted by the AMS.

The director of the dairy division of the AMS will then recommend that a hearing notice be issued once he or she is satisfied that a proposed order is feasible and that those who requested the order can present evidence for the need of the order.

The purpose of the public hearing is to receive information regarding the proposed order from the proponents of the order, representatives of producers, affected handlers, and consumers. The hearing is conducted by a USDA administrative law judge. The hearing is held in the territory of the proposed order in order to encourage participation by those likely to be affected by the order. Witnesses then testify at the hearing and may be questioned. A record of the hearing is then produced.

At the conclusion of the hearings, the administrative law judge sets a time period within which to receive written briefs from the public that would state the court's opinion and interpretation of the hearing process. At the conclusion of this process, the administrative law judge turns the record of the hearings over to the dairy division for recommendation to the secretary of agriculture.

After carefully analyzing the record of the hearings, a recommended decision and tentative order is then issued to the public for comment by the administrator of the AMS. This provides the public with an opportunity to review the decision and to make suggested revisions prior to the final recommendation. After considering all suggested revisions, the AMS then drafts a final decision and forwards it to the secretary of agriculture for review, approval, and issuance.

Producers must then approve the order by referendum. A referendum is a procedure of referring measures proposed by the USDA to dairy farmers to get their approval or disapproval. For orders with a marketwide pool, the order must be approved by at least two-thirds of the producers in the referendum. Once approved by the required percentage of producers, the order is then issued by the secretary of agriculture.

Producers in a proposed order do not have to vote individually in the referendum as the AMAA of 1937 allows for the secretary to accept the vote of a bona fide cooperative as the vote of all its members. This is called "bloc voting." It was justified at the time on the basis of concern that proprietary handlers could attempt to coerce individual producers from voting against a proposed order. In addition, it was felt that bloc voting enabled members of a cooperative to have a unified voice on matters of vital importance to them. Bloc voting is controversial today as some producers feel that their concerns may not be reflected in the cooperative vote.

The procedure for amending an existing order is essentially the same as that for creating a new one. The procedures may differ with regard to the number of days of advance notice required before a hearing or the percentage of producers that must vote to approve the change in a referendum.

REFERENCES

American Agricultural Economics Association Task Force on Dairy Marketing Orders. 1986. Federal Milk Marketing Orders: A Review of Research on Their Economic Consequences. Occasional Paper No. 3, June.

Cropp, Bob and Ed Jesse. 1987. *The Minnesota–Wisconsin Price Series: Its Validity and Alternatives.* Marketing and Policy Briefing Paper 18. University of Wisconsin, September.

Federal Milk Order Study Committee. 1962. *Report to the Secretary of Agriculture.* Washington, D.C.: U.S. Government Printing Office, April.

Hutt, Peter Barton. 1960. "Restrictions on the Free Movement of Fluid Milk under Federal Milk Marketing Orders." *University of Detroit Law Journal,* April.

Manchester, Alden C. 1983. *The Public Role in the Dairy Economy: Why and How Governments Intervene in the Milk Business.* Westview Special Studies in Agricultural Science and Policy. Boulder: Westview Press.

McDowell, Howard, Ann M. Fleming, and Richard F. Fallert. 1988. *Federal Milk Marketing Orders: An Analysis of Alternative Policies.* Agricultural Economic Report 598. Economic Research Service, U.S. Department of Agriculture, September.

McMenamin, Michael and Walter McNamara. 1980. *Milking the Public: Political Scandals of the Dairy Lobby from L.B.J. to Jimmy Carter.* Chicago: Nelson-Hall.

Nicholson, Donald. 1965. "Milk Supplies as Related to Federal Milk Marketing Orders." Ph.D. thesis, Cornell University.

_____. 1994. *A Balancing Act.* Marketing Service Bulletin. Southwest Plains Federal Order, U.S. Department of Agriculture, July 26.

U.S. Department of Agriculture. 1954. *Report of the Federal Milk Order Study Committee on Its Review of the Federal Milk Marketing Order Program.* Chaired by E. W. Gaumnitz. October.

_____. 1964. "Lehigh Valley Cooperative Farmers, Inc. et al. v. United States et al." In *Agriculture Decisions 61–63,* vol. 21, pp. 721–47.

_____. 1972. *Milk Pricing Policy and Procedures: Part I, The Milk Pricing Problem*. Report of the Milk Pricing Advisory Committee. March.

_____. Agricultural Adjustment Administration. 1933. *Marketing Agreement and License for Milk: St. Louis Production Area*. Marketing Agreement Series, Agreement 24; License Series, License 18. Washington, D.C.: U.S. Government Printing Office.

U.S. Department of Agriculture. Agricultural Marketing Service. 1989. *The Federal Milk Marketing Order Program*. Marketing Bulletin 27. Washington, D.C., January.

_____. 1990. *Compilation of Agricultural Marketing Agreement Act of 1937 with Amendments as of January 1, 1990, and Selected Related Legislation*. Washington, D.C., August.

_____. 1991. *Study of Alternatives to Minnesota-Wisconsin Price*. Washington, D.C., September.

_____. 1993. *Southern Illinois-Eastern Missouri Federal Order 1032*. Maryland Heights: Market Administrator Office. July 1.

_____. 1994a. *Recommended Decision*—New England, et al. Lon Hatamiya, administrator. Washington, D.C., August 3.

_____. 1994b. *Amplified Final Decision—National Hearing*. Secretary of Agriculture Mike Espy. Washington, D.C., August 10.

_____. 1994c. *Dairy Market News*. Vol. 61, Report 36.

U.S. Department of Justice. 1977. *Milk Marketing. A Report of the U.S. Justice Department to the Task Group on Antitrust Immunities*. Washington, D.C.: U.S. Government Printing Office, January.

U.S. Department of Agriculture. National Agricultural Statistics Service. 1992. *Statement of the National Agricultural Statistics Service*. U.S. Department of Agriculture Hearing on Alternatives to the Minnesota-Wisconsin Manufacturing Grade Milk Price Series, June 15.

U.S. General Accounting Office. 1988. *Milk Marketing Orders: Options for Change*. GAO/RCED-88-9. Washington, D.C.: Superintendent of Documents, March.

Chapter 7

Dairy Cooperatives

Cooperatives have a long and distinguished history in American agriculture. While they have their roots in northern Europe, they have become an important part of the U.S. economic fabric and are reflective of the U.S. democratic process. While theories abound regarding their form and function, the reasons for their creation are clear and are discussed in this chapter.

Prior to the formation of cooperatives, American farmers had little bargaining power with proprietary firms that processed raw agricultural products or supplied necessary inputs. Despite their large number, farmers felt disadvantaged relative to the railroads, grain mills, and dairy processors, who were few in number but large in size. As a result of this imbalance in the marketplace, cooperatives were formed to represent the needs of many small farmers and to offer a countervailing force in the marketplace.

This chapter covers the formation, development, and functions of dairy cooperatives because of their significant impact on how milk is priced and marketed in the United States. Dairy cooperatives are intimately involved in the creation, amendment, and dissolution of federal milk marketing orders. They also have marketing options that proprietary firms do not have, including the ability to collectively bargain for higher prices on behalf of their members by forming federations, marketing agencies in common, and other mutual arrangements. Dairy cooperatives also provide many services to the marketplace, including balancing the supply of milk between Class I needs and surplus uses.

Definition and Principles of Cooperatives

There are many ways to define a cooperative. One definition quoted by Rasmussen (1991) is "a user-owned and controlled business from which benefits are derived and distributed equitably on the basis of use." Barton notes that cooperatives are distinguished from other businesses by three concepts or principles. First, the user-owner principle: persons who own and finance the cooperative are those who use it. Second, the user-control principle: control of the cooperative belongs to those who use it. Third, the user-benefits principle: benefits of the cooperative are distributed to its users on the basis of their use.

There are four major types of agricultural cooperatives in existence today: marketing, purchasing, service, and mixed-type (Rasmussen 1991). Modern dairy cooperatives are typically classified as marketing cooperatives, although many also have purchasing and service functions as well. Dairy cooperatives ensure a stable market for a highly perishable product, bargain for higher prices on behalf of members, and may offer members patronage dividends. Many dairy cooperatives today manufacture and market processed dairy products and some bottle milk as well.

Some of the earliest known cooperatives in the United States were butter and cheese cooperatives. Rasmussen notes that Ann Pickett of Lake Mills, Wisconsin, organized the earliest reported cheese cooperative in 1841. She and her neighbors supplied the milk, and Mrs. Pickett and her son produced the cheese. Thereafter, growth in cooperatives surged, and by 1890 there were 1,000 agricultural cooperatives in the United States, fully two-thirds of which marketed dairy products.

Dairy cooperatives continued to grow in the early 1900s and began to account for a significant portion of manufactured dairy products. By the 1930s, cooperatives accounted for 36 percent of all U.S. butter production and 25 percent of all cheese production (Manchester 1983b). In addition to manufactured dairy products, cooperatives were also involved in the marketing of fluid milk products. At the turn of the century, many cooperatives were formed as bargaining associations on behalf of dairy farmers. These associations attempted to bargain for higher fluid milk prices and to enforce classified pricing schemes on behalf of their members. Manchester (1983b) reports that by 1927 there were 159 cooperatives that marketed 11 billion pounds of bulk whole milk.

History of Cooperative Legislation

Knapp (1973) outlines the development and growth of American agricultural cooperatives during the period 1920–45. He considered 1920 as a hinge year in the development of cooperatives because it marked the end of postwar agricultural prosperity. By 1921, American farmers were facing the worst depression they had ever known due to a tremendous drop in demand for agricultural products. Knapp argues that three significant developments in 1920 had a dominant influence on the development of cooperatives: (1) the onset of the post–World War I agricultural depression, (2) the emergence of the commodity cooperative marketing movement, and (3) the formation of the American Farm Bureau Federation.

Thus the history of agricultural cooperatives in the United States reveals intrinsic American attitudes toward prosperity and depression, competition and free trade. This brief history lesson begins with a discussion of U.S. antitrust policy and follows with an outline of cooperative legislation.

U.S. Antitrust Legislation

The core foundation of U.S. antitrust legislation consists of (1) the Sherman Act of 1890, (2) the Clayton Act of 1914, and (3) the Federal Trade Commission Act of 1914. At the heart of these acts was Congress's desire to preserve free trade and atomistic competition as the norm for American business as well as its disdain for monopolies and other impediments to free trade.

The Sherman Act of 1890 was rooted in the American industrial revolution that occurred just after the Civil War. Congress became increasingly concerned in the latter part of the nineteenth century that large capital-intensive firms were forming monopolies and thus thwarting free market competition. McBride (1986) notes that the combination of new industrial technologies, market expansion made possible by the railroads, changes in the banking industry that provided commercial risk capital, and liberal state incorporation laws all contributed to growth in the size of American business firms. These firms had tremendous fixed costs due to large investments in capital and therefore reacted to business depressions by either merging with former competitors or driving others out of business.

One anticompetitive technique used by large firms at the time was a legal device called a "trust." *American Jurisprudence* defines a trust as "any combination, whether of producers or vendors of a commodity, for the purpose of controlling prices and suppressing competition, so that all contracts, agreements, and schemes whereby those who are competitors combine to regulate prices." Trusts were legal arrangements whereby the voting stock of different companies were brought together under the direction of a board of trustees in exchange for trust certificates. They were used to consolidate firms and to form a monopoly in order to take over an industry.

The first trust in the United States was established by John D. Rockefeller in 1882. He persuaded the stockholders of 40 companies associated with Standard Oil Company of Ohio to turn over their common stock to nine trustees in exchange for what were then called "trust certificates." What unfolded next was a near monopolization of the U.S. oil industry. Other examples of large companies that effectively used trusts during this period were the United States Steel Corporation, American Sugar Refining Company, and United States Rubber Company.

The Sherman Act of 1890, otherwise known as the Sherman Antitrust Act, was created out of public pressure to thwart the predatory activities of trusts that were driving small firms out of business. Section 1 of the act states, "Every contract, combination in the form of trust or otherwise, or conspiracy, in restraint of trade or commerce among the several States, or with foreign nations, is hereby declared to be illegal." It should be clear that the Sherman Act was not explicitly concerned with the size or even dominance of a firm, but rather with prohibiting use of power to control prices or to exclude competitors from the market (USDA, Capper-Volstead Study Committee 1979).

The Sherman Antitrust Act, however, made no specific reference to cooperatives. In fact, McBride (1986) notes that Congress did consider an amendment to earlier bills that included language specifically related to labor and farm organizations. This amendment, however, never made it into the final bill that became the Sherman Act.

After the passage of the Sherman Act, attempts were made to declare farmer cooperatives illegal through court action (Ingalsbe and Groves 1989). The Sherman Act left the issue of cooperatives unclear just at a time when agricultural cooperatives were beginning to grow in size and to market agricultural products on behalf of their members. The courts were most concerned with the collective actions of cooperatives regarding pricing agreements and other terms of trade. Rasmussen (1991) notes that dairymen during this time were under indictment in Chicago, Cleveland, and St. Paul for selling milk cooperatively in violation of the Sherman Antitrust Act.

The Clayton Act of 1914 was passed to strengthen the Sherman Act by further preventing collusion and other anticompetitive activities that resulted in the formation of monopolies. The Clayton Act also had the objective of clarifying the role of farm cooperatives. Section 6 of the act states, "Nothing contained in the antitrust laws shall be construed to forbid the existence and operation of labor, agricultural or horticultural organizations instituted for the purposes of mutual help and not having capital stock or conducted for profit, or to forbid or restrain individual members of such organizations from lawfully carrying out the legitimate objects thereof." While the Clayton Act did clarify the existence of cooperatives, it did not help those cooperatives at the time that were formed on a capital stock basis. A bigger problem was that it did not deal with the nonprofit criteria since agriculture cooperatives are not nonprofit organizations. After the Clayton Act was passed the Justice Department initiated several actions against cooperatives that prompted Congress to limit their activities through appropriation bills starting in 1914 (Manchester 1983a). Despite this many agricultural cooperatives were found in violation of antitrust laws.

A final component of U.S. antitrust legislation was the Federal Trade Commission Act of 1914. While its formation was not directly related to agriculture or cooperatives, it has in intervening years closely scrutinized the marketing activities of dairy cooperatives, particularly in the 1970s. The Federal Trade Commission (FTC) was created under the Federal Trade Commission Act of 1914 as an independent agency of the U.S. government. Its objective is to promote free and fair trade by investigating and prosecuting unfair trade practices. The FTC is run by five commissioners who are appointed by the president for seven-year terms with the advice and consent of the Senate. An essential component of the mission of the FTC that is of concern to dairy cooperatives is that the FTC has the power to investigate and prevent price-fixing arrangements and to prohibit mergers. While the FTC is charged with enforcing the Sherman and Clayton acts, it has been limited by Congress in its investigation of cooperatives. The USDA has sole authority under the Capper-Volstead Act to supervise agricultural cooperatives.

Cooperative Legislation

The post–World War I depression, beginning in 1920, had a devastating impact on the U.S. farm economy. The following events acted in unison to raise farm production costs and lower farm prices: a higher rediscount rate by the Federal Reserve Board, the elimination of the War Finance Corporation, which financed exports to war-torn Europe, and the termination of government control of railroads. As a result, the index of prices received by farmers declined from 234 in June 1920, when the agricultural depression began, to 115 by the end of 1921 (Knapp 1973). These events had a significant and urgent impact on the presidential administrations of Woodrow Wilson (Democrat, 1913–21), Warren G. Harding (Republican, 1921–23), and Calvin Coolidge (Republican, 1923–29). Knapp notes two important political forces in this period: the inception of the cooperative commodity marketing movement and the McNary-Haugen movement.

The cooperative commodity movement began with ideas spawned by Aaron Sapiro, a young lawyer from California who at the time had witnessed specialty crop marketing in his home state and later became involved in the development of a wheat pool in the Pacific Northwest. Sapiro thought that the cooperative marketing of commodities could work nationwide. Sapiro advocated what he called "cooperation—American style," in contrast to "cooperation—English style," a reference to the Rochdale plan (Knapp 1973). He effectively argued that the Rochdale system, which originated in England in 1844, was for consumer purposes, whereas his system was more directed to a producer cooperative marketing plan.

The essential features of the Sapiro plan as restated by Rasmussen (1991) are as follows:

1. organize on a commodity basis,
2. restrict membership to farmers,
3. employ democratic control of the cooperative by members,
4. maintain long-term, iron-clad contracts with members,
5. organize a large market share before contracts become effective,
6. pool products according to grade,
7. use professional experts in management and other technical positions with the cooperative,
8. adopt sound merchandising principles and techniques,

9. organize on a centralized basis with direct membership,
10. legally incorporate on a nonstock association basis, and
11. maintain orderly marketing.

These 11 principles formed the basis of a movement that started the cooperative marketing of agricultural products in the United States. Many of these principles are still embodied by U.S. dairy cooperatives today.

The American Farm Bureau Federation, formed in 1920, took notice of Sapiro's ideas and saw commodity cooperative marketing as a possible solution to the farm depression. It invited Sapiro to address the Grain Marketing Conference held July 23–24, 1920, in Chicago. The Farm Bureau was impressed with Sapiro's ideas and began to consider the effectiveness of cooperative marketing for other commodities such as cotton and tobacco.

The Harding administration came into power on March 1921 and ushered in a more conservative approach to the farm depression. Harding's new secretary of agriculture, Henry C. Wallace, was formerly the editor of the influential magazine *Wallaces' Farmer* and was prominent in the newly formed American Farm Bureau Federation. It was under Wallace and his chief economist, Henry C. Taylor, that the USDA refocused attention on the need for research of and assistance to cooperatives. Wallace also pressed President Harding to host a national conference of agricultural leaders in order to improve the economic conditions of farmers. His National Agricultural Conference was held January 23–27, 1922, at the Willard Hotel in Washington, D.C. President Harding addressed the conference and emphasized the self-help nature of agricultural cooperatives, "American farmers are asking for, and it should be possible to afford them, ample provision of law under which they may carry on in cooperative fashion those business operations which lend themselves to that method, and which, thus handled, would bring advantage to both the farmer and his consuming public" (Knapp 1973). An important outcome of the conference was strong support for new legislation to further the progress of agricultural marketing cooperatives.

Another idea that came into being in the early 1920s was direct government intervention in the marketplace as a means to stem the farm depression. Equity for agriculture was a concept that grew out of the National Agricultural Conference. George N. Peek, president of the Moline Plow Company, advocated a fair exchange value for all agricultural products. This was the forerunner to the concept of parity pricing, which became part of agricultural legislation in the early 1930s. Peek and his associate General Hugh S. Johnson advocated the development of a two-price discriminatory pricing scheme. The concept was to dump surplus agricultural products onto the world market for whatever they would bring and to limit the domestic portion in order to drive up domestic prices to parity levels. The mechanics of this program were later worked out by Wallace and Taylor. Taylor even suggested to Wallace the creation of an agricultural export commission with broad powers to buy, store, or export farm products and to levy taxes on specific products to pay the costs of stabilizing the products' prices (Knapp 1973). These ideas were then introduced into Congress by Senator McNary of Oregon and Representative Haugen of Iowa every year between 1924 and 1928 (Cochrane and Runge 1992). These bills and the ideas they represented became known as the McNary-Haugen plan, and while they passed both houses of Congress in 1927 and again in 1928, they were vetoed each time by President Coolidge.

It is interesting to note that these same concepts were resurrected in the predebate to the 1995 Farm Bill. The National Milk Producers Federation proposed a self-help plan to form a producer-funded export marketing program to dump surplus dairy commodities onto the world market. The objective of the plan was to short the domestic market and to drive up prices. The self-help plan was effectively abandoned by the National Milk Producers Federation after the November 1994 elections that swept in Republican control of both the House and Senate.

The McNary-Haugen bills were vetoed by President Coolidge because they were thought to involve too much government intervention in the agricultural economy. Support for cooperatives, on the other hand, was viewed as a more feasible alternative. It was from this environment that two major pieces of cooperative legislation were formed: the Capper-Volstead Act of 1922 and the Cooperative Marketing Act of 1926.

The Capper-Volstead Act of 1922, often referred to as the Magna Carta of cooperatives, clarified antitrust law regarding agricultural cooperatives. According to the act, "Persons engaged in the production of agricultural products ... may act together in associations, corporate or otherwise, with or without capital stock, in collectively processing, preparing for market, handling, and marketing in interstate and foreign commerce, such products of persons so engaged." This legislation had its origins in a bill originally drafted by the Dairymen's League, supported by the National Cooperative Milk Producers Federation (forerunner to the National Milk Producers Federation), and later redrafted by Senator Arthur Capper of Kansas and Congressman Andrew Volstead of Minnesota.

Section 1 of the Capper-Volstead Act essentially provides an exemption to agricultural cooperatives from antitrust litigation. Agricultural cooperatives, however, had to operate for the mutual benefit of their members and to conform to the following:

1. That no member of the association is allowed more than one vote because of the amount of stock or membership capital he may own therein.
2. That the association does not pay dividends on stock or membership capital in excess of 8 percent per year.
3. That the association shall not deal in the products of nonmembers to an amount greater in value than such as are handled by its own members.

The act requires number 1 and/or 2 for an organization to be considered an agricultural cooperative. Section 2 of the act requires USDA oversight of all cooperatives in order to ensure that no cooperative monopolizes or restrains trade to such an extent that the price of any agricultural product is unduly enhanced. This section essentially moved the supervisory authority for cooperatives from the Justice Department and the Federal Trade Commission to the Department of Agriculture.

Rasmussen (1991) notes that Section 2 of the act came out of floor debate on the bill regarding the possibility that an agricultural cooperative could become so strong that it would stifle competition, establish a monopoly, and thus unduly enhance prices. Senator Walsh of the Senate Judiciary Committee was clearly worried about cooperatives gaining monopoly power, particularly with highly localized commodities such as fluid milk.

> Your committee sees no good reason why two, three, a half dozen, or a dozen cooperative associations might not properly be organized for the purpose of supplying a city with its milk. ... Section 2 of the House bill has for its purpose the alleviation of the evils of monopoly, or those that may follow from authorized combinations, by the granting of

something like supervisory control to the Secretary of Agriculture. (Manchester 1983a)

The Capper-Volstead Act should not be misinterpreted as exempting agricultural cooperatives from antitrust laws. Cooperative actions that constitute monopolistic behavior are still subject to Justice Department action. In addition, while the Capper-Volstead Act does not limit the size of cooperatives (they can become the sole seller in a given market), they are limited from using monopolization and restraint of trade to unduly enhance prices (USDA, Capper-Volstead Study Committee 1979).

The final major legislation affecting cooperatives was the Cooperative Marketing Act of 1926. Up until that time, USDA support for cooperatives was sporadic and depended in large measure on who was the secretary of agriculture. The Cooperative Marketing Act effectively broadened and formalized USDA support and encouragement of cooperatives (Ingalsbe and Groves 1989). It essentially directed the secretary of agriculture to establish a Division of Cooperative Marketing that would carry on a comprehensive program of service, research, and education for agricultural cooperatives (Rasmussen 1991). This Division of Cooperative Marketing is the forerunner of the Agricultural Cooperative Service.

Advantages of Cooperatives in Milk Marketing

Dairy cooperatives have certain advantages over proprietary firms and therefore have a direct impact on the pricing and marketing of milk in the United States. Some of these advantages accrue from partial exemption from antitrust laws that allow cooperatives to collectively bargain and thereby have the market power to bid up the price of milk. In addition, the creation of federal milk marketing orders has been extremely helpful to cooperatives in bargaining and marketing efforts. Cooperatives have bloc voting privileges that provide them with a large share of votes to create, amend, or terminate federal orders. Also, since a cooperative is considered a single producer under federal orders, it has the ability to pay its members milk prices below minimum federal order prices. Cooperatives, however, still must pay competitive milk prices if they are to retain members.

Section 1 of the Capper-Volstead Act authorized marketing cooperatives to form marketing agencies in common (MAC), otherwise known as "federations of cooperatives." In addition, Section 5 of the Cooperative Marketing Act of 1926 allows cooperatives to "acquire, exchange, interpret, and disseminate past, present, and prospective ... information by direct exchange between such persons, and/or such associations or federations thereof, and/or by and through a common agent created or selected by them."

Dairy cooperatives have used the Section 1 exemption to bargain collectively and to charge bottlers milk prices above minimum Class I prices in select federal orders. In some cases, dairy cooperatives have formally created a MAC to bargain for milk prices in local markets. A MAC is a separate and legal bargaining cooperative. Individual members of the MAC maintain separate individual cooperatives but can unite under a common pricing strategy. To be effective, MACs must consist of all dairy cooperatives that market fluid milk into specific federal orders. That allows them to legally create bargaining power from which they can set a common pricing strategy and bargain for higher milk prices above minimum federal order prices.

The courts have held that cooperatives do not have to create a formal MAC in order to collectively bargain for higher milk prices. They merely have to form a bargaining agency consisting of members that are themselves cooperatives (Manchester 1982). For example, dairy cooperatives that market milk in the St. Louis area have an informal agency called the Southern Illinois–Eastern Missouri Pooling Agency. This group evolved from three pools that were formed in 1968: the St. Louis–Ozarks Agency, the Southern Illinois Agency, and the Central Illinois Agency. Membership in the current agency has changed over the years but in 1994 included the Associated Milk Producers (Morning Glory and North Central regions), Wisconsin Dairies, Mid-America Dairymen, Prairie Farms, Land O'Lakes, and the National Farmers Organization. The agency marketed milk to Borden, Prairie Farms, Pevely, Mid-States Dairy, Chester Dairy, and Land O'Sun. This pooling agency represents an intercooperative combination. While it lacks the legal cooperative status of a MAC, its actions are legal and have been very effective in bargaining for milk prices above federal order minimum prices.

The higher Class I prices charged by bargaining agencies are known as "cooperative Class I prices" and are reported each month for relevant orders by the Agricultural Marketing Service. The difference between cooperative prices and minimum federal order prices is called "overorder payments."

Overorder payments (also known as "overorder premiums" or "service charges") came into existence in the mid-1950s. They were found in about one-third of all federal orders in the early 1960s and grew to about 65 percent in 1970 (Manchester 1980). Their common use arose out of a market shift to larger regional cooperatives in the late 1960s when cooperatives took on increased market functions and provided greater services. In a practical sense, it became easier to form a bargaining association when there was a limited number of players in a given market.

The effectiveness of MACs and bargaining agencies to charge overorder premiums depends on the ability of their members to agree on a common pricing strategy and method to distribute the proceeds, as well as an ability to limit outside sources of lower-priced milk. A common method used to distribute the proceeds from effective bargaining was to form a superpool and to agree to a common pricing formula. Part of this pricing formula would be to allow for competitive credits. These discounts are issued by some cooperatives to their customers in order to remain competitive in a local market. The pricing formula is important since members of the pool may have alternative levels of Class I utilization. Other marketing costs may be deducted from the pool.

A recent example of a MAC is the Upper Midwest Marketing Agency, otherwise known as UMMA, located in federal order 68. UMMA had been successful in bargaining for overorder payments that averaged 81 cents per hundredweight in 1994. Since not all members of UMMA had a similar Class I utilization, the question facing UMMA was how to agree to distribute those proceeds equitably to each member. Obviously those cooperatives with a higher Class I utilization level would benefit more than those with less. So UMMA members agreed to each keep the first 45 cents of overorder payments during the months of January through July, and 65 cents during August through December, for marketing services performed, and to remit the balance to the UMMA superpool to be paid back to all members on a uniform basis. Overorder premiums have been adversely affected in 1995 due to disagreements between cooperatives and because milk was moving such large distances. It is becoming increasingly tempting for some market participants to underbid overorder premiums with "outside milk."

A number of reasons have been cited to justify the existence and use of overorder payments. The General Accounting Office in a report to the U.S. Senate outlined three commonly cited reasons (U.S. GAO 1990). First, overorder payments compensate cooperatives for some services that were formerly provided by other milk handlers. Examples include performing quality control work, paying dairy farmers, and balancing the market for fluid and manufacturing uses. Second, overorder payments facilitate the transfer of milk from surplus to deficit areas. Additional payments may be required in the case of shipping fluid milk from a surplus to a deficit area in the event that the Class I differential does not accurately cover the cost of transporting milk. Finally, overorder payments compensate for imperfections in formula pricing. The General Accounting Office cites as an example the fact that Class I prices are a function of the Minnesota-Wisconsin (M-W) price series (for milk used for manufacturing purposes) lagged two months. This formula may not accurately reflect current market conditions. In addition, federal order prices are minimum prices only. They do not fully account for all market conditions.

Cooperatives also have special privileges under federal orders that proprietary firms don't have. Jacobson (1978) outlines some of these advantages:

1. The cooperative is entitled to bloc vote for its members. In other words, cooperatives can vote on behalf of their members when creating or amending an order. Since only dairy producers can vote on creating or amending an order, dairy cooperatives, especially those with many members, wield tremendous power in specific orders.
2. The cooperative is entitled to blend or pool the proceeds from the sale of member milk. Cooperatives can reblend pool receipts from more than one pool back to their members after netting out transportation and other costs. They can do this since they are considered a single producer under federal orders and are therefore not required to pay their members the minimum prices that proprietary firms must pay producers.
3. The cooperative may collect proceeds, for its members, from handlers that purchase milk. For example, a bottling plant that is supplied by a cooperative may pay class prices directly to the cooperative and settle up with the order pool based on the plant's actual utilization of classified products. Or, if the plant receives cooperative milk directly

shipped from the farm to the plant, it can directly pay the cooperative the blend price and then compensate the pool for the difference between class prices and the blend price. (Proprietary firms also have the ability to do this.)

4. Members of a cooperative that performs marketing services are exempt from market service charges levied by market administrators to nonmembers. That's because cooperatives test their members' milk (for shipment weights, butterfat levels, somatic cell count, etc.) and provide them with market information.
5. Cooperatives may move or direct milk in a manner not permitted proprietary handlers in some cases. For example, in order for a handler's milk to be eligible to receive the blend price, a certain proportion of the handler's producer milk must be delivered to a pooling plant that bottles milk. Under some orders, that minimum amount is often lower for cooperatives than for proprietary handlers (Masson, Masson, and Harris 1978). This provides cooperatives with greater flexibility in marketing milk.

Structure of Dairy Cooperatives in the United States

Dairy cooperatives, owned and managed by dairy farmers, continue to represent a dominant means to market milk from the cow to the consumer. While the number of dairy cooperatives fell from 1,244 in 1964 to 265 in 1992 (due in large part to mergers), the percentage of the national milk supply marketed by dairy cooperatives grew from 67 to 82 percent over the same time period (Stafford and Roof 1984; Ling and Liebrand 1994).

This trend toward greater participation in cooperatives by dairy farmers is no doubt related to the many services cooperatives provide their members. Jacobson outlines four major services most dairy cooperatives provide their members. They

1. Guarantee their member producers a market.
2. Bargain for the best price terms possible.
3. Assemble and market milk as efficiently as possible.
4. Help achieve higher quality levels in milk coming to market.

Participation by dairy farmers in cooperative milk marketing has varied considerably by region of the country. For example, Table 7.1 reveals that the greatest proportion of milk marketed by cooperatives was in the central states. In that region, the percentage of milk marketed by cooperatives increased from 60 percent in 1957 to 96 percent in 1992. Over this same period the proportion of cooperative marketed milk increased in

Table 7.1. Cooperative share of milk delivered to plants and dealers by volume and region

	Region					
	Northeast	South Atlantic	Central	Mountain	Pacific	United States
			(million lb)			
Share handled by cooperatives						
1957	13,239	3,299	35,538	1,995	3,967	58,038
1964	16,956	4,176	47,812	2,683	5,116	76,743
1973	14,541	4,523	54,333	3,210	6,620	83,227
1980	15,033	5,222	61,671	3,834	9,874	95,634
1987	15,047	5,121	66,469	4,790	14,371	105,798
1992	15,624	6,936	71,618	5,271	23,173	122,622
U.S. total delivered to plants and dealers						
1957	19,783	6,584	59,277	3,473	9,261	98,378
1964	22,636	7,489	69,140	3,980	10,690	113,935
1973	20,337	8,198	63,605	4,620	13,063	109,823
1980	23,439	9,301	68,927	5,864	17,186	124,717
1987	25,740	9,279	73,877	7,535	22,661	139,082
1992	26,143	9,805	74,345	10,332	28,179	148,804
			(percent)			
Cooperative percentage of region or U.S. total						
1957	67	50	60	57	43	59
1964	75	56	69	67	48	67
1973	72	55	85	69	51	7
1980	64	56	89	65	57	77
1987	58	55	90	64	63	76
1992	60	70	96	51	82	82

Source: Agricultural Cooperative Service, USDA.

the Pacific region from 43 to 82 percent, and increased from 50 to 70 percent in the South Atlantic region. Cooperative marketed milk actually declined in the Northeast and Mountain states, however, over the period 1980–92.

Of the 265 dairy cooperatives in operation in 1992, a large majority were located in three major regions of the United States (Table 7.2). The Middle Atlantic region (New York, New Jersey, and Pennsylvania) was the largest with 90 cooperatives. The second largest region, the West North Central region (North Dakota, South Dakota, Nebraska, Kansas, Minnesota, Iowa, and Missouri) had 72 cooperatives. The third largest region was the East North Central region (Wisconsin, Michigan, Illinois, Indiana, and Ohio), which had 53 dairy cooperatives.

A little over half of all dairy cooperatives in 1992 did not physically handle dairy products and served as bargaining organizations for their members' milk (Table 7.3). Thirty-two percent of all dairy cooperatives in 1992, most of which were located in the North Central regions, processed and manufactured dairy products. The balance of cooperatives operated milk-receiving facilities only, and most of these were located in the West North Central region.

The top 49 dairy cooperatives in 1994 as surveyed by *Hoard's Dairyman* marketed over 111 billion pounds of milk (Table 7.3). The top 5 cooperatives accounted for 37 percent of the total volume of milk marketed by these 49 dairy cooperatives and 27 percent of all milk marketed in 1994. The number of members in each cooperative, however, was not necessarily correlated with the volume of milk marketed or the size of the farms. For example, Mid-America Dairymen of Springfield, Missouri, was ranked second in the survey and marketed 8.85 billion pounds of milk from 12,819 members for an average of 690,381 pounds per member. The California Milk Producers Association of Artesia, California, on the other hand, ranked third in the survey and marketed 6.11 billion pounds of milk from just 393 members for an average of 15,547,073 pounds per member. The average member of the California Milk Producers Association marketed 22.5 times more milk than the average member in Mid-America Dairymen.

The trend toward fewer and larger dairy cooperatives is clearly illustrated in the *Hoard's Dairyman* survey. There were a number of mergers, consolidations, and acquisitions underway during the last half of 1994 and the beginning of 1995 that were not reported in the *Hoard's Dairyman*

Table 7.2. Dairy cooperatives by type of operation and by headquarters region

	Processing and manufacturing dairy products			Operating milk receiving facilities only			Not physically handling dairy products			Total		
	1980	1987	1992	1980	1987	1992	1980	1987	1992	1980	1987	1992
						(number)						
New England	7	6	3	1	0	0	4	2	2	12	8	5
Middle Atlantic	15	10	9	5	4	6	90	78	75	110	92	90
East North Central	93	55	35	7	3	1	18	17	17	118	75	53
West North Central	42	24	19	77	35	36	5	18	17	124	77	72
South Atlantic	9	6	4	0	1	0	7	5	8	16	12	12
South Central	3	3	3	4	1	1	6	3	3	13	7	7
Mountain	9	5	4	1	0	0	5	2	3	15	7	7
Pacific	14	12	9	2	0	0	11	6	10	27	18	19
All regions	192	121	86	97	44	44	146	131	135	435	296	265
						(percent)						
Percentage of total cooperatives	44	41	32	22	15	17	34	44	51	100	100	100

Source: Agricultural Cooperative Service, USDA.

Table 7.3. Top 49 dairy cooperatives in the United States by volume, 1994

Rank	Dairy cooperative	Member milk volume (billion lb)	Number of members	Rank	Dairy cooperative	Member milk volume (billion lb)	Number of members
1	Associated Milk Producers, Inc. San Antonio, Texas	15.70	13,403	11[a]	California Gold Dairy Products Petaluma, California	3.08	460
2	Mid-America Dairymen Springfield, Missouri	8.85	12,819	11[a]	Manitowoc Milk Producers Cooperative Manitowoc, Wisconsin	3.08	3,449
3	California Milk Producers Assn. Artesia, California	6.11	393	13	Dairylea Cooperative East Syracuse, New York	2.97	2,883
4	Farmers Union Milk Marketing Cooperative Madison, Wisconsin	5.47	10,385	14	Michigan Milk Producers Assn. Novi, Michigan	2.95	2,700
5	Darigold Farms Seattle, Washington	5.00	1,200	15	Western Dairymen Coop Thornton, Colorado	2.84	981
6	Land O'Lakes Minneapolis, Minnesota	4.07	4,950	16	Wisconsin Dairies Cooperative Baraboo, Wisconsin	2.70	4,477
7	Milk Marketing, Inc. Strongsville, Ohio	3.81	5,855	17	Agri-Mark Lawrence, Massachusetts	2.51	2,064
8	Dairymen, Inc. Louisville, Kentucky	3.60	3,159	18	Maryland and Virgina Milk Producers Association Reston, Virginia	2.20	1,289
9	Atlantic Dairy Cooperative Southampton, Pennsylvania	3.58	3,522	19	Florida Dairy Farmers Assn. Fort Lauderdale, Florida	1.91	177
10	Dairymen's Cooperative Creamery Association Tulare, California	3.50	287	20	Southern Milk Sales San Antonio, Texas	1.89	1,234

Table 7.3. (*continued*)

Rank	Dairy cooperative	Member milk volume (billion lb)	Number of members
21	San Joaquin Valley Dairymen Los Banos, California	1.84	227
22	Milwaukee Coop Milk Producers Brookfield, Wisconsin	1.83	1,957
23	Eastern Milk Producers Coop Assn. Syracuse, New York	1.70	2,250
24	United Dairymen of Arizona Tempe, Arizona	1.66	105
25	Danish Creamery Association Fresno, California	1.57	120
26	Allied Federated Cooperative Canton, New York	1.31	1,323
27	Swiss Valley Farms Davenport, Iowa	1.29	2,185
28	Alto Dairy Cooperative Waupun, Wisconsin	1.19	1,279
29	Carolina Virginia Milk Producers Assn. Charlotte, North Carolina	1.04	438
30	Independent Co-op Milk Producers Association Grand Rapids, Michigan	0.99	785
31	Upstate Milk Cooperative LeRoy, New York	0.94	558
32	First District Association Litchfield, Minnesota	0.90	1,200
33	Golden Guernsey Dairy Cooperative Waukesha, Wisconsin	0.87	965
34	Tampa Independent Farmers' Assn. Tampa, Florida	0.85	130
35	Prairie Farms Dairy Carlinville, Illinois	0.78	747
36	St. Albans Cooperative Creamery St. Albans, Vermont	0.77	572
37	Magic Valley Quality Milk Producers Jerome, Idaho	0.70	74
38	Gulf Dairy Association Kentwood, Louisiana	0.62	583
39	Security Milk Producers Ontario, California	0.58	19
40	Tillamook County Creamery Assn. Tillamook, Oregon	0.47	187
41	Cass Clay Creamery Fargo, North Dakota	0.46	836
42	Valley of Virginia Cooperative Milk Producers Association Mount Crawford, Virginia	0.43	293

Table 7.3. (*continued*)

Rank	Dairy cooperative	Member milk volume (billion lb)	Number of members	Rank	Dairy cooperative	Member milk volume (billion lb)	Number of members
43	Cal-West Dairymen Walnut Creek, California	0.39	32	47[a]	Tri-State Milk Cooperative West Salem, Wisconsin	0.28	610
44	Central Pennsylvania Milk Marketing Coop Reedsville, Pennsylvania	0.39	241	49[a]	Cooperative Milk Producers Blackstone, Virginia	0.24	120
45	Farmers Cooperative Creamery McMinnville, Oregon	0.39	108	49[a]	Humboldt Coop Creamery Assn. Fortuna, California	0.24	130
46	Huntington Interstate Milk Producers Huntington, West Virginia	0.30	500	49[a]	Plainview Milk Products Coop Plainview, Minnesota	0.24	286
47[a]	Ellsworth Cooperative Creamery Ellsworth, Wisconsin	0.28	475		Total	111.36	95,022

Source: *Hoard's Dairyman*, October 10, 1994, p. 688,
[a]= tie.

survey. For example, a number of cooperatives merged into the number 2–ranked Mid-America Dairymen of Springfield, Missouri, including the number 8–ranked Dairymen, Inc., of Louisville, Kentucky, the State Dairy Association of Corona, California, the number 38–ranked Gulf Dairy Association of Kentwood, Louisiana, the number 20–ranked Southern Milk Sales of San Antonio, Texas, and the Coble Dairy of Lexington, North Carolina. As of January 1995, Mid-America Dairymen had 16,702 members in 30 states ranging from California to Georgia and from Wisconsin to Texas. These Mid-America Dairymen mergers will likely raise its rank to number 1 in member milk volume in 1995, ahead of the Associated Milk Producers.

In addition to the Mid-America Dairymen mergers, Allen Dairy Products of Fort Wayne, Indiana, merged into number 35–ranked Prairie Farms of Carlinville, Illinois. Also, number 16–ranked Wisconsin Dairies Cooperative of Baraboo, Wisconsin, merged with number 33–ranked Golden Guernsey Dairy Cooperative of Waukesha, Wisconsin, to form Foremost Farms USA. Foremost Farms USA then purchased the assets and assumed the liabilities of the Morning Glory Division of the Associated Milk Producers. Milk Marketing Inc. of Strongsville, Ohio, merged with Eastern Milk of Syracuse, New York. While the cooperative is still called Milk Marketing, a study committee was formed to come up with a new name.

This new rash of mergers, which in essence are forming national dairy cooperatives, is reminiscent of the mergers in the late 1960s and early 1970s that formed regional cooperatives. The U.S. Justice Department reported that from 1968 to 1972, 217 local cooperatives had merged into four regional cooperatives: Associated Milk Producers, Inc. (AMPI), Mid-America Dairymen, Inc. (Mid-Am), Dairymen, Inc. (DI), and Milk, Inc. (U.S. Department of Justice 1977). AMPI was incorporated on October 1, 1969. It was formed from the Pure Milk Association and Milk Producers and covered the Midwest and Southwest. Mid-Am was effectively formed on July 1, 1968, through a series of mergers with 31 cooperatives serving Iowa, Kansas, Missouri, and Illinois (Reeves 1989). DI, located in the Southeast, was incorporated two months after Mid-Am. DI was formed from a consolidation of 8 cooperatives that then merged with 16 other cooperatives in its first three years of existence. Milk, which was later renamed Milk Marketing, was originally incorporated from a consolidation of four cooperatives in Ohio and Pittsburgh.

Dairy cooperatives today are involved in all aspects of milk marketing, such as the traditional roles of bargaining for milk prices on behalf of their members, receiving and assembling milk, overseeing its quality, and hauling it to processors. They are also involved in the manufacture and processing of various dairy products. Dairy cooperatives are under pressure from their members to market more value-added products. In 1992, dairy cooperatives accounted for 65 percent of all butter manufactured in the United States, 43 percent of all cheese, 48 percent of all dry whey products, and 81 percent of all dry milk products (Table 7.4). They were less involved in the manufacture and processing of Class I and Class II products such as packaged fluid milk, cottage cheese, yogurt, ice cream, and ice cream mixes.

Antitrust Actions and Undue Price Enhancement

Growth in the size of dairy cooperatives in the late 1960s, combined with inflation concerns in the early 1970s, focused intense scrutiny on the pricing policies of major cooperatives. At issue was whether or not dairy

Table 7.4. Percentage of U.S. dairy products distributed by cooperatives

	1980	1987	1992
		(percent)	
Packaged fluid milk products	16	14	16
Cheese	47	45	43
Butter	64	71	65
Dry milk products[a]	87	91	81
Bulk condensed milk	15	45	27
Cottage cheese	22	13	1
Dry whey products	81	53	48
Yogurt[b]	NA	NA	3
Ice cream mix and ice milk mix	7	11	13
Ice cream and ice milk	11	8	10

Source: USDA, Agricultural Cooperative Service.

Note: NA = data not available.

[a]Includes nonfat dry milk, dry whole milk, and dry buttermilk.

[b]Includes plain and fruit-flavored yogurt and frozen yogurt.

cooperatives were forming monopolies and thereby restraining trade. Overorder payments were becoming more commonplace in many federal orders. As a result, the Justice Department, the Federal Trade Commission, the USDA, and Congress all became involved in examining potential monopoly powers of cooperatives in relation to antitrust legislation and the undue price enhancement provisions of the Capper-Volstead Act.

At the time Congress became concerned that, once formed, mergers of any kind of company that led to monopolies were extremely difficult to break when assets, management, technology, and marketing systems were consolidated. As a result, Congress passed the Hart-Scott-Rodino Antitrust Improvements Act of 1976. Under this act, companies that intend to merge and meet certain conditions must prenotify the Justice Department and provide specific economic data about the merger. Once notification is made, the Justice Department has 30 days in which to respond. If the notification form reveals problem areas, the Justice Department can request more specific information. The Justice Department then has 20 days thereafter to respond.

The Hart-Scott-Rodino Act was not passed to challenge every merger. The intent was to apply this law annually to the (approximately) largest 150 mergers, as these are the most likely to substantially lessen competition (House report). Mid-America Dairymen, for example, had to notify the Justice Department prior to its merger with Dairymen, Inc.

The Justice Department and the Federal Trade Commission became concerned about the monopoly power of large firms in the early 1970s. It was during this period that they filed suit against American Telephone and Telegraph (AT&T) for monopoly power over the telecommunications industry. The Justice Department and the Federal Trade Commission became interested in dairy cooperatives on two grounds: first, due to concerns that recent merger activities would lead to monopolies and therefore restraint of trade and, second, because cooperatives could collectively bargain and charge overorder payments.

The Antitrust Division of the Justice Department launched three antitrust suits against the three major dairy cooperatives: AMPI, Mid-Am, and DI. They were charged with attempted monopolization and restraint of trade. Both AMPI and Mid-Am entered into consent decrees with the Justice Department. A consent decree is essentially an agreement reached between the parties of the case to agree to certain actions. After it is approved by a judge, the case can be dismissed. DI, however, fought the case and eventually prevailed in 1976.

These cases were well publicized. A major report was issued by Justice Department economists outlining the case against AMPI. The report concluded that the structure, conduct, and performance of AMPI substantiated the existence of a monopoly. It was particularly critical of cooperative conduct patterns including overorder premiums, the standby pool, full-supply contracts, and cooperative pooling practices (Eisenstat et al. 1971).

The standby pool refers to the Associated Reserve Standby Pool Cooperative (ARSPC), which was a federation of Upper Midwest and southern milk marketing cooperatives formed in 1970. The objective of ARSPC was to maintain a reserve supply of milk for bottling plants in the South. This in effect made northern milk supplies available to southern markets at reasonable prices. The Justice Department maintained, however, that the ARSPC was used to charge overorder premiums to southern processing plants by keeping northern milk supplies out of southern federal order pools.

Full-supply contracts were agreements between bottlers and cooperatives whereby the cooperative would supply all of the fluid needs of the bottler. To some, full-supply contracts were developed to meet the very special marketing needs of fluid milk processors. But to others, they were used by some cooperatives to lock out alternative supplies of milk to a bottling plant. If the bottler bought a portion of its milk from a competing cooperative or nonmember producers, then that would violate the terms of the contract, and milk shipments by the contracting cooperative would be terminated. This strategy helped cooperatives bargain for higher prices since bottlers were concerned about losing an important supply of fluid milk.

In 1976 the Justice Department report on AMPI was reviewed and critiqued by three university economists: Hugh Cook, Leo Blakley, and Calvin Berry. They were critical of the report and offered alternative justifications for the use of overorder premiums, the reserve standby pool, and full-supply contracts. Their criticism, however, appeared almost derogatory, as evidenced in the following statement, "These [alleged major oversights] arose no doubt in part from their [Justice Department economists'] lack of familiarity with the economics of agriculture, the peculiarities of fluid milk and of its production and handling, and with the institutions within which dairy farmers, cooperatives and handlers operate as well as the history which gave rise to those institutions." Cook, Blakley, and Berry reached the following conclusions, "Given their conclusions that AMPI was a monopoly, Masson and associates [the Justice Department] covered and discussed most of the reasons why a firm such as AMPI could not perform as a monopoly over time. Their report is more complete because they included discussion of why monopoly gains for AMPI and its members could not be maintained even if achieved in the short run."

The U.S. Department of Agriculture was petitioned a number of times in the late 1960s and early 1970s to investigate cases of undue price enhancement on the part of cooperatives. While it was the intent of the Capper-Volstead Act to allow farmers to cooperatively market their products and thus enhance prices, they could not go too far (Manchester 1983a). Section 2 of the Capper-Volstead Act requires the secretary of agriculture to take legal action against any agricultural cooperative that is responsible for undue price enhancement through monopolization or restraint of trade. What constitutes undue price enhancement is left to interpretation by the courts, like all antitrust terms.

Three inquiries resulted from complaints from milk buyers in 1969, 1972, and 1973 (Manchester 1980). A fourth case resulted from the Judiciary Committee of the House of Representatives in 1973. Another case originated from a petition filed by the National Consumers Congress. It alleges that the price of milk was unduly enhanced in any federal order market where overorder premiums exceeded 50 cents per hundredweight in any month in 1975.

The USDA responded by forming the Capper-Volstead Study Committee to examine the role of the department in enforcing Section 2 of the Capper-Volstead Act (USDA, Capper-Volstead Study Committee 1979). That committee reported in December 1976 that there had not been any cooperatives found guilty of under-price enhancement in any court case. Therefore undue price enhancement is still undefined by the courts. The committee defined a standard for judging undue price enhancement as "whether or not a farmer cooperative has gone beyond the level of equality in market power in negotiating price and trade terms. Prices that significantly exceed the level associated with equality of market power would constitute undue price enhancement" (USDA, Capper-Volstead Study Committee 1979). The committee's point was that it was the legislative intent of the Capper-Volstead Act to enhance farmer prices to a certain extent by providing farmers with greater market power equal to that of other firms. Beyond that level, a level not defined in the act, undue price enhancement would result.

The role and effects of Section 2 of the Capper-Volstead Act were recently examined by the General Accounting Office in a report to the Subcommittee on Antitrust, Monopolies, and Business Rights, of the Senate Committee on the Judiciary, and Senator Bill Bradley (U.S. GAO 1990). The GAO concluded that (1) technological improvements in milk production and transportation have helped the individual dairy farmer's relative market strength over what it was in the 1920s when the Capper-Volstead Act was passed, (2) research on cooperative pricing power is inconclusive, and (3) the USDA was not actively monitoring cooperative pricing activities. The GAO concluded that if the USDA does not initiate active monitoring of cooperative activities, Congress should consider assigning regulatory responsibility for such matters to the Federal Trade Commission.

It should be mentioned that there is no evidence to date that dairy cooperatives are practicing undue price enhancement. The ability of farmers to disassociate themselves from a cooperative and to bargain directly with a processor has kept overorder premiums in check.

REFERENCES

American Jurisprudence. 1941. Vol. 36, *Monopolies, Combinations, and Restraints of Trade*, §19. Rochester: The Lawyers Co-operative Publishing Co.

Barton, David. 1989. "What Is a Cooperative?" In *Cooperatives in Agriculture*, edited by David Cobia. New Jersey: Prentice-Hall.

Cochrane, Willard and C. Ford Runge. 1992. *Reforming Farm Policy: Toward a National Agenda*. Ames: Iowa State University Press.

Cook, Hugh L., Leo Blakley, and Calvin Berry. 1976. *Review of Eisenstat, Philip, Robert T. Masson, and David Roddy "An Economic Analysis of the Associated Milk Producers, Inc., Monopoly."* R2790 Research Bulletin. Research Division, College of Agricultural and Life Sciences, University of Wisconsin-Madison, January.

Cook, Hugh L., Leo Blakley, Robert Jacobson, Ronald Knutson, Robert Milligan, and Robert Strain. 1978. *The Dairy Subsector of American Agriculture: Organization and Vertical Coordination*. North Central Regional Research Project NC 117, Monograph 5.

Eisenstat, Philip, Robert T. Masson, and David Roddy. 1971. "An Economic Analysis of the Associated Milk Producers, Inc., Monopoly." U.S. Department of Justice, Washington, D.C.

Hart-Scott-Rodino Antitrust Improvement Act 90 Stat. 1383 (1976).

H.R. Report 1376, 94th Cong., 2nd sess. 11 (1976), reprinted in 1976 USCCAN 2637, 2643.

Ingalsbe, Gene and Frank Groves. 1989. "Historical Development." In *Cooperatives in Agriculture,* edited by David Copia. New Jersey: Prentice-Hall.

Jacobson, Robert E. 1978. "Sources, Limits and Extent of Cooperative Market Power: Cooperatives and Marketing Orders." *Agricultural Cooperatives and the Public Interest.* Proceedings of a North Central Regional Research Committee 117–sponsored workshop. NCR Research Publication 256. September.

Knapp, Joseph G. 1973. *The Advance of American Cooperative Enterprise: 1920–45.* Danville: Interstate Printers and Publishers.

Ling, K. Charles and Carolyn Betts Liebrand. 1994. *Marketing Operations of Dairy Cooperatives.* ACS Research Report 133. Agricultural Cooperative Service, U.S. Department of Agriculture, April.

McBride, Glynn. 1986. *Agricultural Cooperatives: Their Why and Their How.* Westport: AVI Publishing Company.

Manchester, Alden C. 1980. "Viewpoint of the U.S. Department of Agriculture." *Proceedings of the 24th National Conference of Bargaining and Marketing Cooperatives.* ESCS-83. Economics, Statistics, and Cooperative Service, U.S. Department of Agriculture. Washington, D.C.: National Technical Information Service, July.

_____. 1982. "The Status of Marketing Cooperatives under Antitrust Law." ERS-673. National Economics Division, Economic Research Service, U.S. Department of Agriculture, February.

_____. 1983a. "Agricultural Marketing Cooperatives and Antitrust Laws." *Antitrust Treatment of Agricultural Marketing Cooperatives.* North Central Regional Research Publication 286. NCR Project 117, Monograph 15. Studies of the Organization and Control of the U.S. Food System, September.

_____. 1983b. *The Public Role in the Dairy Economy: Why and How Governments Intervene in the Milk Business.* Westview Special Studies in Agricultural Science and Policy. Boulder: Westview Press.

Masson, Alison, Robert T. Masson, and Barry C. Harris. 1978. "Cooperatives and Marketing Orders." *Agricultural Cooperatives and the Public Interest.* Proceedings of a North Central Regional Research Committee 117–sponsored workshop, September. NCR Research Publication 256.

Rasmussen, Wayne D. 1991. *Farmers, Cooperatives, and USDA: A History of Agricultural Cooperative Service.* Agricultural Information Bulletin 621. Washington, D.C., July.

Reeves, James (Jim) L. 1989. *The First 20 Years: The Story of Mid-America Dairymen.* Republic: Western Printing Company.

Stafford, Thomas H. and James B. Roof. 1984. *Marketing Operations of Dairy Cooperatives.* ACS Research Report 40. Agricultural Cooperative Service, U.S. Department of Agriculture, July.

_____. Capper-Volstead Study Committee. 1979. "Undue Price Enhancement by Agricultural Cooperatives: Criteria, Monitoring, Enforcement." June.

U.S. Department of Agriculture. 1989. *Questions and Answers on Federal Milk Marketing Orders.* AMS-559. Dairy Division, Agricultural Marketing Service, Washington, D.C., September.

U.S. Department of Justice. 1977. *Milk Marketing.* A Report of the U.S. Justice Department to the Task Group on Antitrust Immunities. Washington, D.C.: U.S. Government Printing Office, January.

U.S. General Accounting Office. 1990. *Dairy Cooperatives: Role and Effects of the Capper-Volstead Antitrust Exemption.* GAO/RCED-90-186. Washington, D.C.: Superintendent of Documents, September.

Chapter 8

Federal Price Support Program

Government intervention in the U.S. marketplace began in the early 1930s in response to the massive collapse of the U.S. economy during the Great Depression. One objective of government programs then was to support farm prices and farm income. In the case of dairy, this was accomplished by either loans to processors, or more commonly, through direct purchases of manufactured dairy products. This would in effect support the farm price of milk at predescribed levels. The support programs were later modified during World War II in order to encourage an expansion in production. Permanent legislation that created the price support program known today followed in 1949.

Since the inception of the dairy price support program, the farm price of milk was linked to parity prices. That essentially prevented prices from collapsing in some years (i.e., following World War II and the Korean War) and expanded production in other years through higher support prices (i.e., during the war years and the 1960s). Prices were supported by government purchases of unlimited quantities of surplus dairy products (mainly cheese, butter, nonfat dry milk, and evaporated milk). Domestic and foreign donation and other feeding programs were created in order to dispose of these surplus dairy products.

The basic philosophy that has prevailed since the early 1930s was that direct government intervention in the marketplace was needed in order to equate supply with demand at a predetermined farm price. Unfortunately, that has also meant in some years significant government purchases of dairy products and large budgetary outlays (i.e., 1979–89).

History of the Price Support Program

Rojko (1957) reports that the first experiment with price stabilization of a dairy product occurred on January 9, 1934, when, after a recommendation by the Dairy Advisory Committee, the Federal Farm Board granted a loan to Land O'Lakes Creameries. The purpose of the loan was to enable the cooperative to temporarily withhold its own butter from the market and to purchase additional supplies if necessary to stabilize butter prices. While the cooperative was not able to purchase any butter from the market with offers for 35 cents per pound, it did accumulate 5 million pounds of its own production, which was sold back to the market by May of that year at a profit. Thus began government intervention in the marketplace to support and stabilize prices.

Price Support Programs during the Depression

Government actions during the Depression years were in response to the dramatic fall in farm milk prices due to the collapse in the purchasing power of consumers.[1] The precipitous drop in the index of farm prices (Figure 8.1) between 1929 and 1932 would be comparable to a drop in today's prices from $13.00 per hundredweight to $6.50. Not many of us can imagine milk prices tumbling that much.

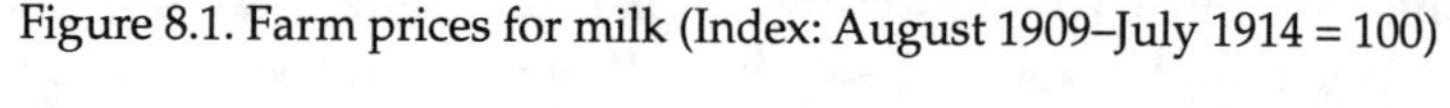

Figure 8.1. Farm prices for milk (Index: August 1909–July 1914 = 100)

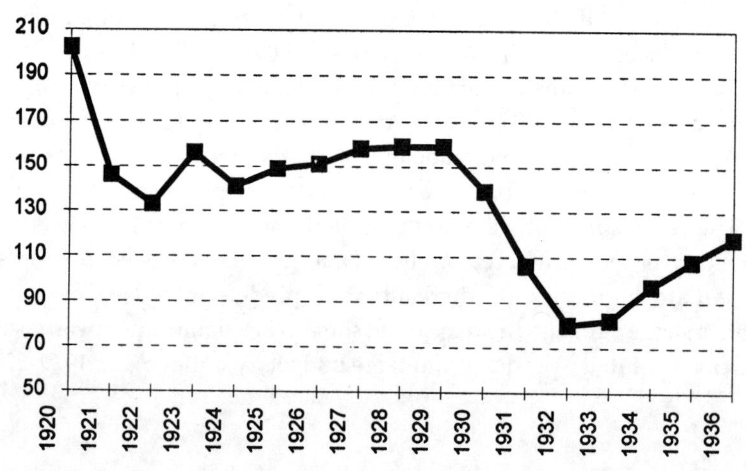

8/Federal Price Support Program

The Agricultural Adjustment Act (AAA) of 1933 (P.L. 73-10, May 12, 1933) was enacted to restore farm purchasing power to parity levels. It provided three ways to improve prices and income to dairy farmers: first, by designating milk and its products as basic commodities that were entitled to price support and production-adjustment operations; second, by authorizing marketing agreements, licenses, and the secretary of agriculture's orders (Chapter 6); and, third, by making funds available to the secretary of agriculture to expand markets and dispose of surplus agricultural products.

A dairy-adjustment program to reduce the supply of milk was never adopted due to the opinion commonly held that production control was difficult if not impracticable for dairy products (Black 1935). Restricting production did not seem feasible to large numbers of dairy farmers at the time. The AAA of 1933 did, however, create the first price support program. According to Black, "The butter purchase program of the Administration is important primarily from the standpoint of the light it throws upon the possibilities of price control through buying and subsequently disposing of 'surpluses.'"

Purchases were made from regular market channels on the basis of competitive bids and did not exceed what could be used for school lunch, institutional, and welfare purposes (U.S. GAO 1990). These purchase programs were carried out with wide discretionary powers from the Agricultural Adjustment Administration and the Federal Emergency Relief Administration. It was under this and subsequent legislation that support for milk was carried out between mid-1933 through 1941 (Table 8.1).

Funds were made available for the purchase of surplus dairy products in 1934 from Sections 2 and 6 of the Jones-Connolly Act, and in the following year from Sections 32 and 37 of the Agricultural Adjustment Act of 1935 (P.L. 74-320, August 24, 1935). Additional funds were made available from various federal and state emergency relief funds. These funds were used to purchase surplus quantities of butter, cheese, evaporated milk, and nonfat dry milk (Table 8.1). These products were then used for relief purposes by the Federal Surplus Relief Corporation (chartered in 1933) under the direction of the Federal Emergency Relief Administration. The Federal Surplus Relief Corporation was later renamed the Federal Surplus Commodities Corporation in 1935 and was transferred to the USDA.

Table 8.1. Dairy products: Purchases by the USDA, mainly for price support, 1933–41

				Whole milk equivalent, all purchases		Nonfat dry milk		
							Purchases as % of —	
Year	Butter	Cheese	Evaporated milk	Quantity purchased	Purchases as % of milk production on farms	Quantity purchased	Production of nonfat dry milk for human use	Total solids-not-fat produced on farms
	(1,000 lb)	*(1,000 lb)*	*(1,000 lb)*	*(million lb)*	*(%)*	*(1,000 lb)*	*(%)*	*(%)*
1933	43,234	—	—	869	0.8	—	—	—
1934	24,624	17,936	400	675	0.7	—	—	—
1935	7,055	192	47,027	244	0.2	15,840	8.4	0.2
1936	2,951	932	6,160	82	0.1	3,594	1.6	—
1937	3,049	138	19,636	104	0.1	23,188	9.5	0.3
1938	141,979	3,463	19,470	2,916	2.8	31,260	10.8	0.3
1939	25,398	—	3,209	515	0.5	5,035	1.9	0.1
1940	10,604	4,354	65,903	397	0.4	7,317	2.3	0.1
1941	11,454	—	4,350	238	0.2	2,742	0.7	—

Source: Rojko 1957.

The Dairy Products Marketing Association was created in 1938 as a nonprofit organization consisting of eight regional butter cooperatives to operate the government stabilization program for butter. It received loans from the Commodity Credit Corporation, or CCC (chartered in 1933), to buy butter at specified prices. This butter was to be held in storage for possible resale through normal commercial channels at prices sufficient to cover the purchase price plus handling and carrying charges. Dairy products not sold back to the market could be sold to the Surplus Marketing Administration for relief distribution. This was the precursor to today's dairy price support program.

Price Support Programs during World War II

The emphasis of price support programs shifted during World War II from removing surplus dairy products from the market to supporting farm prices to expand production for wartime uses and maintaining domestic prices below price ceilings. Demand for both manufactured dairy products for U.S. troops and allies and fluid milk at home increased significantly during the war years. To help assist dairy farmers to meet these needs, the government increased dairy support prices and used various subsidy programs in order to maintain price ceilings for consumers.

The USDA began to purchase substantial quantities of American cheese, evaporated milk, and nonfat dry milk for wartime shipment to the British (U.S. GAO 1990). Price supports were used to encourage milk production to meet these needs. On April 1, 1941, the USDA announced that it would support prices of dairy products through June 30, 1943, by market purchases of butter in Chicago for 31 cents per pound.

The Steagall Amendment (P.L. 41-147, July 1, 1941) further enhanced the price support program by requiring milk to be supported at no less than 85 percent of parity. In October 1942, Congress amended the Emergency Price Control Act and raised the minimum support levels to 90 percent of parity. These support levels were to be maintained at such levels for two years beyond the cessation of hostilities. Thus support for manufactured dairy products such as evaporated milk, nonfat dry milk, cheese, and butterfat became mandatory for the first time and were linked to parity price levels.

The CCC took over responsibilities for purchasing and distributing dairy products for the price support program when the Steagall amendment was enacted. Market purchases to support milk and dairy prices, however, were not necessary due to strong wartime demand. Large purchases by the armed forces and the USDA for military, lend-lease, and postwar foreign assistance programs kept market prices during and immediately after the war above support levels.

In addition to the price support program and direct market purchases, subsidy programs were developed during the war years to improve farmer returns in order to expand production yet keep domestic prices at or below price ceilings established by the Office of Price Administration (Henderson 1955). The CCC paid subsidies directly to dairy producers, cheddar cheese processors, and milk handlers. These payments were made as compensation for high production costs due to the war and to keep consumer prices below price ceilings. For example, Congress initiated a program in October 1943 to subsidize dairy farmers for feed grain purchases. Government-imposed price ceilings for corn and a high price support level for hogs encouraged on-farm consumption of feed grains, which limited their availability to dairy farmers. These subsidy payments were continued until price ceilings were terminated on July 1, 1946 (U.S. GAO 1990).

In addition to these subsidies, the Defense Supplies Corporation, a wartime subsidiary of the Reconstruction Finance Corporation, made direct payments to creameries in the amount of 5 cents per pound of butter from June 1943 to October 1945. That was to allow those creameries to absorb a 5-cent price rollback ordered by the Office of Price Administration.

Postwar Price Support Programs

During the war years, legislators and farm leaders were concerned about the possibility of a farm depression developing immediately following World War II like that following the first world war. As a result, two actions were taken (Cochrane and Runge 1992). First, administrators of war food programs kept a "bare shelf" food stocking policy. Second, Congress passed the Steagall amendment, which provided mandatory price support for dairy products at 90 percent of parity through December 31, 1948.

Unfortunately, these wartime policies led to problems meeting the needs of war-torn Europe and Asia immediately following the war. As a result, farm and retail prices soared between 1945 and 1948. As for the dairy price support program under the Steagall amendment, only 211 million pounds of nonfat dry milk were purchased in 1947.

Policy debate heated up following World War II as Congress searched for new directions for farm policy. Cochrane and Runge report that two camps evolved. The first camp was characteristically Republican and wanted to reduce government intervention in agriculture by lowering the level of price support on farm commodities. The second camp was characteristically Democratic party leaders from the South and the plains. They wanted to maintain high price supports as a means to protect farm prices and income.

The second camp clearly won the debate as the Agricultural Act of 1948 extended mandatory price supports at 90 percent of parity through 1949. In the following year, Congress permanently adopted the price support programs created during the war years to support farm income. The Agricultural Act of 1949 (P.L. 81-439, October 31, 1949) required the secretary of agriculture to support the prices received by farmers for milk and butterfat at 75 to 90 percent of parity in order to "ensure an adequate supply of milk." This legislation as amended has provided the basic legislation for the price support program from January 1, 1950, to date.

The Agricultural Act of 1949 and subsequent amendments provided a minimum and maximum support price equal to 75 and 90 percent of parity, respectively. The secretary of agriculture was authorized to determine the specific support price within this range that ensured an adequate supply of milk. The support price was to be announced before the start of the marketing year. While the support price couldn't be decreased during the year, it could be increased up to 90 percent of parity.

Section 201 of the act required that support be carried out through loans or direct market purchases of milk and the products of milk and butterfat. With the exception of a limited number of nonrecourse loans made to manufacturers of whey products in 1954, support for milk has been achieved through purchase programs operated by the CCC. Each year, prior to the beginning of the marketing year, the secretary of agriculture announced that the CCC stood ready to buy all quantities of butter, cheddar cheese, and nonfat dry milk at specified prices for specified grades and sizes. These specified prices—called "CCC purchase prices"—were linked by a predetermined formula to the support price for manufacturing grade milk.

From a practical standpoint, support for farm prices can only reasonably be carried out through market purchases of manufactured dairy products since fresh milk is bulky and perishable. Manufactured dairy products can be sold by the trade to the CCC whenever commodity prices fall to support levels, and then repurchased later when prices strengthen. Products sold to the CCC are considered surplus dairy products since market prices would have fallen below the support price had they not been removed from the market. Thus in many years, the price support program had set a market price that was well above market-clearing levels. The support price for milk represents a targeted floor since the actual value of manufacturing grade milk could in some years fall below the milk support price level (e.g., the 1980s).

The dairy price support program also effectively supports prices of manufactured dairy products other than butter, nonfat dry milk, and cheddar cheese (e.g., whey and buttermilk) since all dairy products have a close price relationship. In addition, the market price of milk used for fluid and other Class I and Class II uses is also effectively supported since this price is linked via formulas to the manufacturing grade price for milk (Class III and Class IIIa prices) under federal milk marketing orders (Chapter 6). Thus a clear link exists between the price support program and all federal milk marketing orders.

The CCC purchase prices for butter, nonfat dry milk, and cheddar cheese were determined via formulas linked to the support price for milk. They were set at a level at which a butter/nonfat dry milk plant and a cheese plant would be willing to offer farmers at least the support price for milk. In addition, CCC purchase prices were set in order to ensure a similar level of support to both a butter/powder plant and a cheese plant.

The USDA has had the authority in previous farm legislation to adjust the relationship between the CCC purchase prices of nonfat dry milk and butter so long as it maintained the announced support price. For example, the support price during the 1991–92 marketing was $10.10 for manufacturing grade milk testing 3.67 percent butterfat. During that same marketing year the USDA lowered the CCC purchase price for butter twice, from 98.25 cents per pound to 87.25 cents per pound on January 17, 1992, and then again to 76.25 cents per pound on May 13,

1992. In order to maintain a constant support price of $10.10, however, the CCC purchase price for nonfat dry milk (nonfortified in 50-pound bags) was raised twice, from 85 cents per pound to 91.20 cents per pound on January 17, 1992, and to 97.30 cents per pound on May 13, 1992.

The change in the CCC purchase price of butter is inversely related to the change in the CCC purchase price of nonfat dry milk. This inverse relationship is commonly referred to as the "butter/powder tilt."

Changes in Price Support Legislation, 1970–90

Throughout the 1950s and early 1960s production exceeded consumption, and surplus dairy products were purchased under the dairy price support program in the amount of 3 to 9 percent of total marketings (USDA 1968). The exception was in 1951–52 during the Korean War. Milk production sharply declined in the mid-1960s due to an increased exodus of farmers out of dairying. Farmers were confronted with unfavorable economic conditions resulting from rising production and living costs and unsatisfactory milk prices. That reduced the supply of manufacturing grade milk, lowered production of butter and nonfat dry milk, and reduced CCC purchases.

The secretary of agriculture responded by increasing the support price on April 1, 1966, from $3.24 per hundredweight to $3.50. Support prices were again raised the same year on June 30 to $4 per hundredweight, or 90 percent of parity. In addition, the secretary announced other actions to raise Class I prices under federal orders. By the 1970s, production inputs such as feed, fertilizer, and petroleum products increased rapidly, further exacerbating a price-cost squeeze. Congress responded with the Food and Agriculture Act of 1977 (P.L. 95-113, September 29, 1977), which set a minimum support price of 80 percent of parity. In addition, support prices would be adjusted semiannually to reflect prices paid by farmers. While these support price provisions were to only last two years, they were extended another two years in 1979. As a result, support prices increased beyond the rate of inflation during the 1970s and early 1980s (Figure 8.2). The support price, for example, increased 46 percent between 1977 and 1980, compared with an overall inflation rate of 27 percent.

Higher support rates fueled an expansion in milk production well beyond domestic needs. Milk production expanded from a low of 115 bil-

Figure 8.2. Minnesota-Wisconsin price series for milk and the support price for milk

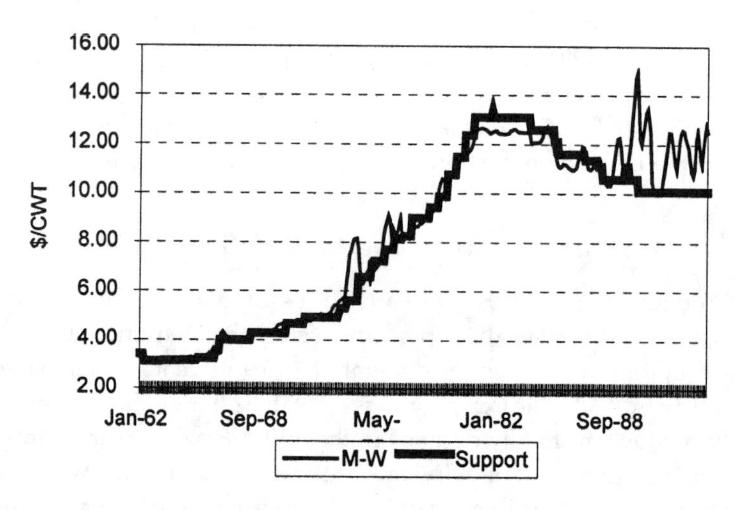

lion pounds in 1975 to 128 billion in 1980 (Figure 8.3). That equates to an annual average increase of 2.3 percent. Since the commercial market was unable to buy that much milk at or above support price levels, CCC purchases and the cost of the price support program escalated rapidly. CCC removals of American cheese, for example, increased from 49 million pounds in 1970 to 350 million pounds in 1980 (Table 8.2). Likewise, nonfat dry milk removals increased from 452 million pounds in 1970 to 634 million pounds in 1980. Total removals of all dairy products under USDA programs as a percentage of marketings grew from 5.3 percent to 7.1 percent on a milkfat basis, and from 5.2 percent to 8.6 percent on a skim solids basis over the same period. Program costs escalated as well. CCC net expenditures for the dairy price support and related programs increased from $384 million in the 1970–71 marketing year to $2 billion in the 1980–81 marketing year (Table 8.3). Clearly something had to be done.

Congress responded by disengaging the milk price support from parity prices. This was first accomplished by freezing the support price at $13.10 per hundredweight effective October 1, 1980. Next Congress established a set of triggers that related the level of the minimum support

Figure 8.3. U.S. milk production

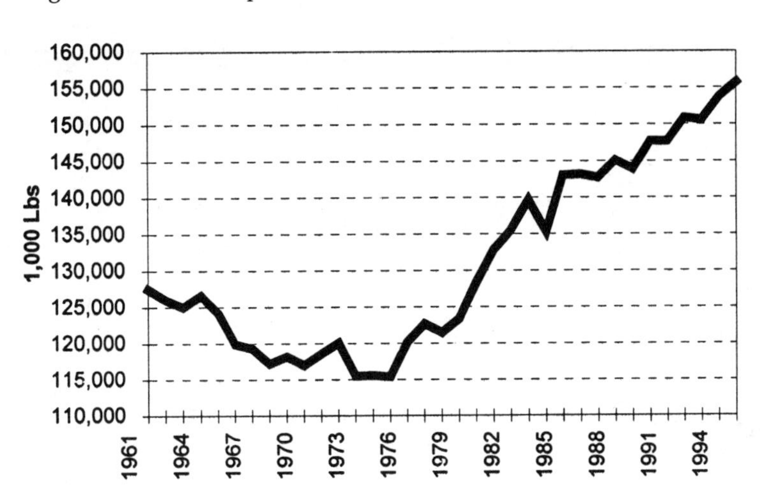

price to the size of CCC purchases under the Agriculture and Food Act of 1981 (P.L. 97-98, December 22, 1981). Support prices could only increase if CCC purchases fell to stated levels. That formally removed the linkage between the price support level and the parity index that existed since the 1930s. The 1981 act also directed the USDA to reduce the CCC's product inventory. Beginning December 1981, the USDA began giving surplus cheese to states for distribution to the needy (U.S. GAO 1990). This program expanded to USDA's Special Distribution Program, which also provided butter and nonfat dry milk to states to distribute to the needy. The Temporary Emergency Food Assistance Act of 1983 (Title II, P.L. 98-8, March 24, 1983) formalized the USDA's program and called it the Temporary Emergency Food Assistance Program (TEFAP). This program directed the USDA to make any CCC commodities such as cheese, butter, and nonfat dry milk that were in excess of quantities needed for other programs and activities available for distribution to the needy. The Hunger Prevention Act of 1988 (P.L. 100-435, September 19, 1988) extended activities under TEFAP through September 1990.

In order to help curb the cost of the price support program, Congress authorized a 50-cent deduction on all milk marketed with the Omnibus Budget Reconciliation Act of 1982. Funds were first collected April 1983.

Table 8.2. Dairy products removed from the commercial market by USDA programs

Year	Butter[a] (million lb)	American cheese (million lb)	Evaporated milk[b] (million lb)	Nonfat dry milk (million lb)	Milk equivalent (milkfat) (million lb)	As a % of marketings (%)	Milk equivalent (skim solids) (million lb)	As a % of marketings (%)
1970	246.4	48.9	48.4	451.6	6,027	5.3	5,845	5.2
1971	292.2	90.7	111.4	456.2	7,547	6.6	6,448	5.6
1972	233.7	30.4	97.0	345.0	5,660	4.9	4,526	3.9
1973	97.7	3.2	53.7	36.8	2,283	2.0	582	0.5
1974	32.7	60.3	28.3	265.0	1,389	1.2	3,728	3.3
1975	63.4	68.2	24.5	394.5	2,151	1.9	5,302	4.7
1976	39.4	38.0	21.8	157.1	1,291	1.1	2,245	1.9
1977	221.8	148.2	15.9	461.7	6,340	5.3	6,874	5.7
1978	112.0	39.7	17.6	285.0	2,909	2.4	3,743	3.2
1979	81.6	40.2	16.4	255.3	2,243	1.9	3,399	2.8
1980	257.0	349.7	17.5	634.3	9,008	7.1	10,875	8.6
1981	351.5	563.0	18.6	851.3	13,087	10.1	15,513	11.9
1982	382.0	642.5	20.8	948.1	14,512	10.9	17,429	13.1
1983	413.2	832.8	24.6	1,061.0	16,982	12.4	20,632	15.0
1984	202.3	447.3	19.0	678.4	8,730	6.6	12,430	9.4
1985	334.2	629.0	26.8	940.6	13,356	9.5	17,216	12.2
1986	287.6	468.4	28.8	827.3	10,837	7.7	14,311	10.2
1987	187.3	282.0	24.1	559.4	6,861	4.9	9,343	6.7
1988	312.6	238.1	23.1	267.5	9,120	6.4	5,540	3.9

Table 8.2. *(continued)*

Year	Butter[a] (million lb)	American cheese (million lb)	Evaporated milk[b] (million lb)	Nonfat dry milk (million lb)	Milk equivalent (milkfat) (million lb)	As a % of marketings (%)	Milk equivalent (skim solids) (million lb)	As a % of marketings (%)
1989	413.4	37.4	28.8	0.0	9,419	6.6	480	0.3
1990	400.3	21.5	30.7	117.8	9,017	6.2	1,689	1.2
1991	442.9	76.9	27.9	269.5	10,425	7.1	3,938	2.7
1992	439.5	14.4	32.8	136.7	9,936	6.6	1,989	1.3
1993[c]	288.8	8.3	25.9	304.3	6,654	4.5	3,876	2.6

Source: USDA 1994.

Note: Figures determined on a delivery basis, after unrestricted domestic sales. Dairy products removed from the commercial market include removals under the Dairy Export Incentive Program and similar export programs and may include purchases under Sections 709 and 4a.

[a]Includes butter equivalent of anhydrous milkfat.

[b]Starting in 1991, no longer considered a price support removal.

[c]Preliminary.

Table 8.3. Parameters of the dairy price support program and related net expenditures

Marketing year (except as noted)	Manufacturing grade milk: Support level: Percentage of parity equivalent (%)[d,e]	Support level: At 3.5% milkfat ($/cwt)	Support level: At national average milkfat test[a] ($/cwt)	Farm price at average milkfat test ($/cwt)	Total milk production (billion lb)	USDA net market removals, milkfat basis (milk equivalent)[b] (billion lb)	Net expenditures on dairy price support and related programs[c] (million $)
1949 (Feb. 8–Dec. 31)	90		3.14	3.14		2.5	188.1
1950–51 (Jan. 1950–Mar. 1951)	81		3.07	3.35	116.2	3.7	(50.0)
April–March marketing year							
1951–52	86	3.23	3.60	3.97	114.3	NA[f]	9.1
1952–53	90	3.49	3.85	4.00	116.5	3.6	300.0
1953–54	89	3.43	3.74	3.46	121.8	11.3	474.4
1954–55	75	2.89	3.15	3.15	121.7	5.1	257.4
1955–56	80	2.89	3.15	3.19	124.4	5.1	284.2
1956–57							
Apr. 1–17	82	2.91	3.15				
Apr. 18–Mar. 31	84	3.00	3.25	3.31	124.7	5.1	331.1
1957–58	82	3.02	3.25	3.28	124.3	6.8	360.0
1958–59	75	2.85	3.06	3.16	123.2	3.5	231.3
1959–60	77	2.86	3.06	3.22	122.7	3.4	218.2
1960–61							
Apr. 1–Sept. 16	76	2.86	3.06				
Sept. 17–Mar. 9	80	3.02	3.22				
Mar. 10–Mar. 31	85	3.20	3.40	3.30	123.2	3.3	281.3
1961–62	83	3.20	3.40	3.38	126.3	11.2	612.0
1962–63	75	2.93	3.11	3.19	125.8	8.8	485.5
1963–64	75	2.99	3.14	3.24	126.2	7.5	379.1

Table 8.3. (*continued*)

Marketing year (except as noted)	Manufacturing grade milk: Support level: Percentage of parity equivalent (%)[d,e]	Manufacturing grade milk: Support level: At 3.5% milkfat ($/cwt)	Manufacturing grade milk: Support level: At national average milkfat test[a] ($/cwt)	Manufacturing grade milk: Farm price at average milkfat test ($/cwt)	Total milk production (billion lb)	USDA net market removals, milkfat basis (milk equivalent)[b] (billion lb)	Net expenditures on dairy price support and related programs[c] (million $)
1964–65	75	3.00	3.15	3.29	126.9	8.2	385.6
1965–66	75	3.09	3.24	3.45	122.2	2.9	229.3
1966–67							
Apr. 1–June 29	78	3.35	3.50				
June 30–Mar. 31	89	3.84	4.00	4.11	119.8	2.7	170.8
1967–68	87	3.85	4.00	4.07	118.2	7.0	417.7
1968–69	89	4.13	4.28	4.30	116.6	4.8	329.6
1969–70	83	4.13	4.28	4.55	116.2	4.4	273.1
1970–71	85	4.51	4.66	4.76	117.4	7.1	383.5
1971–72	85	4.79	4.93	4.91	119.4	6.6	369.9
1972–73							
Apr. 1–Mar. 14	79	4.79	4.93				
Mar. 15–Mar. 31	85	5.17	5.29	5.22	119.1	4.9	264.6
1973–74							
Apr. 1–Aug. 9	75	5.17	5.29				
Aug. 10–Mar. 31	80	5.49	5.61	6.95	114.9	0.7	77.0
1974–75							
Apr. 1–Jan. 4	81	6.45	6.57				
Jan. 5–Mar. 31	89	7.10	7.24	6.87	115.6	2.4	318.7
1975–76							
Apr. 1–Oct. 1	79	7.10	7.24				

Table 8.3. (*continued*)

Marketing year (except as noted)	Manufacturing grade milk: Support level: Percentage of parity equivalent (%)[d,e]	At 3.5% milkfat ($/cwt)	At national average milkfat test[a] ($/cwt)	Farm price at average milkfat test ($/cwt)	Total milk production (billion lb)	USDA net market removals, milkfat basis (milk equivalent)[b] (billion lb)	Net expenditures on dairy price support and related programs[c] (million $)
Oct. 2–Mar. 31	84	7.55	7.71	8.12	116.4	0.9	258.9
1976–77							
Apr. 1–Sept. 30	80	7.95	8.13				
Oct. 1–Mar. 31	81	8.07	8.26	8.52	120.9	3.4	302.4
Oct.–Sept. marketing year							
1976–77[g]							
Oct. 1–Mar. 31	81	8.07	8.26				
Apr. 1–Sept. 30	82 82	8.79	9.00	8.65	122.2	6.9	714.3
1977–78							
Oct. 1–Mar. 31	82	8.79	9.00				
Apr. 1–Sept. 30	86 80	9.21	9.43	9.30	121.6	3.2	451.4
1978–79							
Oct. 1–Mar. 31	80	9.64	9.87				
Apr. 1–Sept. 30	87 78	10.51	10.76	10.86	122.4	1.1	250.6
1979–80							
Oct. 1–Mar. 31	80	11.22	11.49				
Apr. 1–Sept. 30	86 79	12.07	12.36	11.75	127.2	8.2	1,279.8
1980–81	80	12.80	13.10	12.71	131.7	12.7	1,974.8
1981–82[h]							
Oct. 1–Oct. 20		13.18	13.49				
Oct. 21–Sept. 30		12.80	13.10	12.66	134.7	13.8	2,239.2

Table 8.3. (*continued*)

Marketing year (except as noted)	Manufacturing grade milk				Total milk production (billion lb)	USDA net market removals, milkfat basis (milk equivalent)[b] (billion lb)	Net expenditures on dairy price support and related programs[c] (million $)
	Support level						
	Percentage of parity equivalent (%)[d,e]	At 3.5% milkfat ($/cwt)	At national average milkfat test[a] ($/cwt)	Farm price at average milkfat test ($/cwt)			
1982–83		12.80	13.10	12.66	138.7	16.6	2,600.4
1983–84							
Oct. 1–Nov. 30		12.80	13.10				
Dec. 1–Sept. 30		12.31	12.60	12.47	136.6	10.4	1,597.5
1984–85							
Oct. 1–Mar. 31		12.31	12.60				
Apr. 1–June 30		11.81	12.10				
July 1–Sept. 30		11.31	11.60	12.13	140.1	11.5	2,181.4
1985–86		11.31	11.60	11.41	144.8	12.3	2,419.8
1986–87							
Oct. 1–Dec. 31		11.31	11.60				
Jan. 1–Sept. 30		11.07	11.35	11.58	141.5	5.4	1,237.7
1987–88							
Oct. 1–Dec. 31		10.82	11.10				
Jan. 1–Sept. 30		10.33	10.60	11.03	144.8	9.7	1,346.2
1988–89							
Oct. 1–Mar. 31		10.33	10.60				
Apr. 1–June 30		10.83	11.10				
July 1–Sept. 30		10.35	10.60	11.93	144.6	9.6[i]	712.0
1989–90							
Oct. 1–Dec. 31		10.35	10.60				

Table 8.3. (*continued*)

Marketing year (except as noted)	Manufacturing grade milk: Support level: Percentage of parity equivalent (%)[d,e]	Support level: At 3.5% milkfat ($/cwt)	Support level: At national average milkfat test[a] ($/cwt)	Farm price at average milkfat test ($/cwt)	Total milk production (billion lb)	USDA net market removals, milkfat basis (milk equivalent)[b] (billion lb)	Net expenditures on dairy price support and related programs[c] (million $)
Jan. 1–Apr. 21		9.88	10.10				
Apr. 22–Sept. 30		9.90	10.10	13.28	147.0	8.4[i]	523.3
1990–91		9.90	10.10	10.67	148.6	10.4[i]	854.9
1991–92							
Oct. 1–Jan. 16		9.90	10.10				
Jan. 17–May 12		9.94	10.10				
May 13–Sept. 30		9.97	10.10	12.03	150.9	10.2[i]	251.9
1992–93							
Oct. 1–July 6		9.97	10.10				
July 7–Sept. 30		10.00	10.10	11.61	152.0	7.8[i]	268.7
1993–94							
Oct. 1–		10.00	10.10				

Source: USDA, Agricultural Stabilization and Conservation Service.

Note: Dairy Price Support Program and related net expenditures shown as follows: 1949–50 through 1963–64, fiscal year (July 1–June 30); 1964–65 through 1976–77, marketing year (April 1–March 31); 1976–77, 12-month period (October 1–September 30, 1976–77); 1977–78 to date, marketing year (October 1–September 30).

[a]Standardized to 3.67 percent milkfat in 1972–73.

[b]Price support, Section 32, Section 709, and Section 4(a) purchases (delivery basis) of butter, cheese, nonfat dry milk, evaporated milk, and dry whole milk, plus exports on which CCC issued payment-in-kind certificates, less CCC sales for unrestricted

Table 8.3. (*continued*)

Marketing year (except as noted)	Manufacturing grade milk				Total milk production (billion lb)	USDA net market removals, milkfat basis (milk equivalent)[b] (billion lb)	Net expenditures on dairy price support and related programs[c] (million $)
	Support level						
	Percentage of parity equivalent (%)[d,e]	At 3.5% milkfat ($/cwt)	At national average milkfat test[a] ($/cwt)	Farm price at average milkfat test ($/cwt)			

uses.

[c]Net expenditures to CCC plus domestic and foreign donation expenditures. For 1964–65, includes costs of military milk program, and through 1965–66, the payment-in-kind export program. Includes costs of dairy products bought at market prices under Section 709 of the Food and Agriculture Act of 1965 in 1966–67, 1969–70, 1970–71, 1972–73, 1973–74, and 1989–90. Also includes costs of dairy products bought at market prices under Section 4(a) of the Agriculture and Consumer Protection Act of 1973 in 1973–74 and 1974–75.

[d]As of the beginning of the marketing year.

[e]As of April 1.

[f]Less than 50 million pounds.

[g]For 12-month comparative purposes.

[h]Support price established at specific price levels rather than parity levels, beginning October 1981.

[i]Reflects approved methodology for calculating milk equivalent on fat solids basis in accordance with the Food, Agriculture, Conservation, and Trade Act of 1990. Includes removals under the Dairy Export Incentive Program.

A second 50-cent deduction was implemented on September 1, 1983, and was refundable to producers who reduced their marketings by a specified amount.

The Agriculture and Food Act of 1981 and the 50-cent deductions did not stem the growth in milk production and CCC purchases. The cost of the price support program continued to expand and peaked at $2.6 billion for the 1982–83 marketing year (Table 8.3). The Dairy Production Stabilization Act of 1983 (Title I, P.L. 98-180, November 29, 1983) attempted to control supplies by authorizing a new price support trigger mechanism, established the Milk Diversion Program, and created a nationwide Dairy Promotion Program. The 1983 act immediately reduced the support price from $13.10 to $12.60 effective December 1, 1983. It required the secretary of agriculture to reduce the support price 50 cents per hundredweight on April 1 and July 1, 1985, if CCC purchases of dairy products for the 12-month period following those dates exceeded specific levels. The secretary subsequently reduced the support price to $12.10 on April 1 and to $11.60 on July 1, 1985.

The 1981 act also established the Dairy Promotion Program, which was designed to increase consumption of dairy products through promotion, nutrition research, and education activities. The program is funded by a mandatory 15-cents-per-hundredweight checkoff on all milk commercially marketed. Of the total checkoff, 5 cents went to the National Dairy Promotion and Research Board (NDPRB) and the balance to approved state, regional, and local organizations. The United Dairy Industry Association (UDIA), for example, is an approved regional organization funded by the checkoff program. The NDPRB collected the whole 15 cents if there were no other approved organizations.

The Milk Diversion Program became the nation's first voluntary supply management program. The program was temporary in that it was effective between January 1984 through March 1985. Under the program, dairy producers in the 48 contiguous states could contract with the CCC to reduce their marketings by 5 to 30 percent of their milk marketings during a base period (1982 or, at the producer's option, an average of 1981–82 marketings). In return, producers would receive $10.00 for each hundredweight of milk marketing reduction. To help pay for the program, producers were assessed 50 cents per hundredweight.

The General Accounting Office reported that some 38,000 milk producers were paid about $955 million to reduce their milk sales from a base period by 7.5 billion pounds in 1984 and 1.9 billion pounds in the first quarter of 1985 for a total of 9.4 billion pounds (U.S. GAO 1985). This was equivalent to an average reduction of 23 percent of the participants' total milk marketings. The GAO estimated that the program reduced 1984 milk production by 3.74 to 4.11 billion pounds, and reduced CCC purchases by $614 to $664 million. The effects of the program were clearly temporary as milk production in the year following the program (1985) exceeded milk production prior to the diversion program.

The dairy title to the Food Security Act of 1985 (P.L. 99-198, December 23, 1985) continued to emphasize dairy surplus and budgetary control provisions. The act modified the 1983 price support trigger mechanism, authorized the Dairy Termination Program (DTP), increased Class I price differentials for 35 of 44 federal orders, and initiated a dairy export program to target unfair trade practices called the Dairy Export Incentive Program (DEIP).

The 1985 act required the milk support price to be lowered from $11.60 to $11.35 on January 1, 1987, and to $11.10 on October 1, 1987. Thereafter, the secretary of agriculture was required to reduce the support price by 50 cents per hundredweight on January 1 of 1988, 1989, and 1990 if the projected CCC removals of dairy products exceeded 5 billion pounds on a milk-equivalent basis. The support price could be increased by 50 cents per hundredweight on these same dates if projected CCC removals were 2.5 billion pounds or less.

The secretary of agriculture reduced the support price by 50 cents to $10.60 on January 1, 1988. Drought legislation enacted in August 1988 prevented a January 1989 reduction in the support price and required a temporary 50-cents-per-hundredweight support price increase to $11.10 from April through June of 1989. This was to compensate dairy farmers for any higher feed costs as a result of the drought of 1988. The support price was reduced back to $10.60 effective July 1, 1989, and was reduced another 50 cents per hundredweight on January 1, 1990, to $10.10.

The USDA changed the butter/powder tilt during the support price adjustments in order to lower the cost of the price support program. When the milk support price was increased 50 cents per hundredweight on April 1989, the CCC purchase price for butter remained constant at $1.32 per pound and the CCC purchase price for nonfat dry milk was increased from 72.75 cents per pound to 79 cents per pound for 50-pound bags. Likewise, when the support price was lowered 50 cents per hundredweight in July 1989 and January 1990, the CCC purchase price for nonfat dry milk remained constant whereas the CCC purchase prices for butter and cheese were lowered (Table 8.3). These actions lowered the cost of the price support program because relatively more butter was purchased than nonfat dry milk.

The Dairy Termination Program was designed to reduce the size of the nation's dairy herd, and hence milk production and CCC removals. The program allowed producers to voluntarily submit bids to terminate milk production over a period of five years. The cost of this program was partially funded by producer assessments. The bids reflected the payment producers would be willing to accept in return for participation in the program. If accepted, producers had to slaughter or export their dairy cows, replacement heifers, and calves older than 18 months over the period April 1, 1986, to August 31, 1987. In addition, participating producers had to idle their dairy facilities for a period of five years. The cost of this program was partially funded by producer assessments.

The GAO reported that the USDA accepted bids from about 14,000 dairy farmers, which represented about 12.3 billion pounds of milk sales in 1985 (U.S. GAO 1990). Federal payments to participating farmers totaled $1.8 billion. Marsh (1988) reports that dairy cow slaughter increased 6.2 percent in 1986 and 2.2 percent in 1987 over 1985 levels due to the DTP. In addition, 292,503 head of heifers and 271,599 head of calves were slaughtered. The increase in slaughter increased red meat supplies by 410 million pounds in 1986 and 145.5 million pounds in 1987. To alleviate the impact of the DTP on cattle prices, the USDA was authorized to purchase up to 400 million pounds of additional red meat during the liquidation period. The cattle price impact over the 18-month liquidation period ranged from reductions of $1.98 per hundredweight in the steer carcass market and 80 cents per hundredweight in the cow slaughter market. It is debatable whether beef prices were negatively impacted solely by the DTP since increased slaughter of beef cows may have also contributed to the market decline.

The Food, Agriculture, Conservation, and Trade Act (FACTA) of 1990 (P.L. 101- 508, November 5, 1990), passed by Congress in the fall of 1990, continued the market-oriented policies of the 1985 act. The dairy title to FACTA (Title 1) modified the formula for the support price, required the secretary to submit a milk inventory management report to Congress, specified a producer assessment program to pay for surplus government purchases in excess of 7 billion pounds annually, and required the secretary to complete federal hearings and initiate recommendations on a replacement to the Minnesota-Wisconsin (M-W) price series for milk used for manufacturing purposes (Congressional Research Service 1990).

In addition to FACTA, Congress simultaneously passed the Omnibus Budget Reconciliation Act of 1990 (P.L. 101-508, November 5, 1990), which altered the provisions of FACTA to achieve $11.9 billion in budget savings over the five-year life of FACTA. This law had significant effects on the dairy program.

The formula for setting the support price was modified under FACTA and cannot fall below a minimum of $10.10 per hundredweight for milk at 3.67 percent butterfat. The secretary was instructed to estimate CCC purchases of dairy products on November 20 each year for the ensuing calendar year. The estimates were to be calculated on a milk-equivalent basis that included both a milkfat and a solids-not-fat conversion. Prior to FACTA, dairy products purchased under the CCC price support program were converted to milk-equivalent measures on the basis of butterfat conversions only. If the milk equivalent of the projected CCC purchases were to exceed 5 billion pounds, then the secretary was instructed to reduce the support price by 25 to 50 cents per hundredweight effective January 1 of the ensuing year. The support price, however, could not fall below a minimum of $10.10 per hundredweight. If, on the other hand, projected CCC purchases were to fall between 3.5 and 5 billion pounds, no change in the support price was required. If projected purchases were to fall below 3.5 billion pounds, the secretary was to increase the support price by a minimum of 25 cents per hundredweight. The secretary never changed the support price from the level of $10.10 per hundredweight over the life of FACTA. That was because the support price was $10.10 at the beginning of FACTA, the minimum level allowed, and the USDA each year projected CCC purchases to exceed 3.5 billion pounds on a milk-equivalent total solids basis.

The secretary completed an inventory management report and submitted it to Congress on June 14, 1991 (USDA, ASCS 1991). Congress wanted to evaluate the feasibility of government-mandated programs that would reduce domestic milk marketings and raise the farm price of milk. This study was born out of concern by Congress that the market would not be able to reduce milk production on its own and that CCC dairy purchases would someday became excessively burdensome. The secretary analyzed four inventory management options: a target price-deficiency payment program, reclassification (Class IV) program, a two-tier pricing program, and a milk diversion program. The secretary concluded that each program "comes with a number of shortcomings."

As a result of the secretary's recommendation, a milk inventory management program was not adopted by Congress. FACTA required that a milk producer assessment program be adopted in its place (Congressional Research Service 1990). This program required dairy farmers to pay the full cost of any surplus CCC purchases of dairy products in excess of 7 billion pounds in calendar years 1991–95. This would be accomplished in any year via an assessment on all milk marketed during a calendar year. However, since the secretary never projected CCC purchases to be in excess of 7 billion pounds, such an assessment was never required.

The Omnibus Budget Reconciliation Act of 1990 required a separate producer assessment to help defray the cost of the dairy price support program. All dairy producers were assessed 5 cents per hundredweight for all milk marketed in calendar year 1991, and 11.25 cents per hundredweight in calendar years 1992 through 1995. The purpose of the assessment was to reduce the government's budgetary exposure to the price support program. For example, in fiscal year 1994, marketing assessments covered nearly 50 percent of the total cost of the dairy price support program.

The annual assessment was refunded to any producer who did not increase milk marketing from the previous year. When such refunds occurred, however, assessments for all producers would be increased above the base assessment rate of 11.25 cents per hundredweight during the months May through December in the subsequent year in order to recoup the revenue lost from the refund. For example, the assessment rate increased during May through December 1994 from the base rate of 11.25 cents per hundredweight to 19.28 cents per hundredweight after more than $80 million was refunded to producers who did not increase their marketings in 1993.

Like all farm programs, the price support program is an entitlement program. It is in effect an "obligation" on the part of the federal government. The program's legislative authority lays out parameters for how it is to operate (e.g., price support levels, producer assessments, etc.). After the program is authorized, however, it draws funds each year directly from the U.S. Treasury. This contrasts with other programs that need annual appropriations acts. Since the early 1980s, however, entitlement programs have been carefully scrutinized by Congress. Budget reconciliation acts have been used each year to reign in expanding program costs in an overall effort to reduce the size of the growing federal deficit. In fact, the 1990 FACTA was simultaneously amended with budget reconciliation legislation.

The 1996 Federal Agricultural Improvement and Reform (FAIR) Act began the process to phase out the dairy price support program over time. The milk support level was raised from $10.10 per hundredweight in 1995 to $10.35 in 1996. Every year thereafter, it declines by 15 cents per hundredweight to $9.90 in 1999.

The 1996 act also eliminated producer assessments. These assessments, used to partially fund the cost of the dairy price support program, were no longer needed since significant program savings would be realized by the support price reductions.

Net Expenditures and Utilization of Dairy Products

The CCC typically purchases dairy products at those times of the year when there are surplus dairy products on the market. For example, the government may purchase butter and nonfat dry milk during the spring months when milk production seasonally peaks and milk in excess of fluid needs is processed into storable dairy products. These purchases of dairy products from the commercial market by the price support program are termed "removals." The CCC, however, has been authorized to sell back purchased dairy products to the market whenever market prices are at least 110 percent of the original purchase price. These products are referred to as "CCC sellbacks." Net removals essentially consist of CCC purchases minus CCC sales for unrestricted use (CCC sellbacks). The price support program thus purchases surplus dairy products during surplus months of the year, and it sells back some of these products when the market is short.

It should be noted that dairy products sold under the Dairy Export Incentive Program are considered as part of "net removals" by the price support program. The DEIP, discussed in Chapter 10, is used to subsidize exports of dairy products overseas. Although the program assists commercial exports (which is different than foreign donations), it is counted as a government removal from the commercial market. That is because those export sales would not exist were it not for government assistance from the DEIP program.

CCC purchases account for the greatest portion of the price support program costs. Gross outlays for the program include not only the cost of dairy products purchased by the CCC but also the associated costs of transporting, storing, and handling such dairy products. Another large component of gross outlays is the cost of the DEIP. The cost of this program is reflected in gross outlays since it has a budgetary cost and removes surplus dairy products from the market. Another gross outlay is the cost of assessment refunds to producers who did not expand production.

For example, fiscal year 1993 (October 1, 1992, to September 31, 1993) had gross outlays of $679 million. Of that amount, $315 million was for purchases of surplus dairy products, $61.3 million was for storage, handling, transportation, processing and packaging, $134.4 million was for the DEIP program, $50.8 million was for assessment refunds, and the balance ($117.2 million) was for "other expenses or outlays."

Gross outlays, however, do not reflect the final budget cost to taxpayers since additional sources of revenue offset this cost. Such offsetting income, or receipts, consists mainly of the proceeds of CCC sellbacks and producer assessments. For example, in fiscal year 1993, such income consisted of $103.6 million from the proceeds of sales, $202.2 million in producer paid gross assessments, $117.5 million in other income or receipts (mainly CCC export sales at market prices), and $2.9 million in net transfers of purchased dairy products to blended foods (Table 8.4). The total offsetting income, or receipts, for fiscal year 1993 was $426.3 million, which resulted in net CCC outlays of $252.7 million.

The CCC has utilized dairy products purchased under the price support program for many types of programs (USDA 1968). Four major uses are as follows:

- *Sales for unrestricted uses.* CCC sellbacks to the market are examples of sales for unrestricted use. Such sales occur at prices above purchase prices in order to recoup CCC costs for storage and handling. The program has been consistent with congressional policy set forth in Section 407 of the 1949 act for basic and storable commodities.
- *Sales for restricted uses.* These sales are used mainly for out-of-condition

Table 8.4. Net CCC outlays under the dairy price support program

	FY 1990	FY 1991	FY 1992	FY 1993
		(million $)		
Gross outlays				
Purchases	398.9	756.8	394.5	315.0
Storage and handling	16.5	22.5	25.7	21.4
Transportation	13.4	19.9	23.3	16.3
Processing and packaging	9.1	16.9	30.1	23.6
Other expenses or outlays	44.6	60.2	128.2	117.2
Diversion payments	0.0	0.0	(0.1)	0.0
Termination payments	188.8	96.1	2.4	0.0
Dairy Export Incentive Program	0.0	0.0	24.0	134.6
Assessment refunds	0.0	0.0	23.6	50.8
Subtotal	671.4	972.4	652.1	679.0
Receipts				
Proceeds from sales	122.7	25.6	135.6	103.
Net transfers to blended foods	0.0	3.0	10.4	2.9
Other income or receipts	36.2	60.4	127.9	117.5
Milk marketing (assessments) refunds	7.7	44.5	146.3	202.2
Subtotal	166.6	133.5	420.1	426.3
Net CCC outlays	504.8	838.9	232.0	252.7

Source: USDA 1994.
Note: Figures rounded off.

stock (e.g., nonfat dry milk no longer fit for human consumption) for animal feed.

- *Commercial export sales.* These reflect both DEIP program sales as well as government-to-government sales.
- *Domestic donations.* The CCC has used available purchased dairy products in school lunch, welfare, U.S. military, and Veterans Administration hospital programs as well as other domestic donation programs.
- *Foreign donations.* The CCC has also donated available purchased dairy products to foreign governments.

For accounting purposes, the CCC each month releases reports of purchases and stocks of dairy products acquired under the dairy price support program. Available stocks of dairy products that are committed to various domestic feeding programs are accounted for as "committed inventories." Only those inventories considered "uncommitted" are available as sellbacks to the market.

Domestic donation programs under the price support program do not represent all government-assisted feeding programs. The National School Lunch Program, authorized by the National School Lunch Act of 1946, provides healthy meals at subsidized prices to children in participating schools. Part of this program includes packaged fluid milk. Fluid milk is also provided in the Special Milk Program, which services children in child-care centers, summer camps, and similar nonprofit institutions. Another domestic feeding program that uses dairy products is the Women, Infants, and Children (WIC) Program, which provides nutritious food to expectant mothers and their children.

All three programs receive funding outside of the dairy price support program and purchase dairy products either through CCC inventory or from commercial outlets. In recent years, due to the volatility of dairy commodities, the CCC has advised these programs as to which time of the year to purchase dairy products from the market. Since the CCC is given a fixed appropriation and since it must purchase dairy products at market prices, more dairy products are made available to program beneficiaries when market prices are low.

NOTES

1. Rojko (1957) was the primary source of information for this section.

REFERENCES

Black, John D. 1935. *The Dairy Industry and the AAA.* Washington, D.C.: Brookings Institution.

Cochrane, Willard and C. Ford Runge. 1992. *Reforming Farm Policy: Toward a National Agenda.* Ames: Iowa State University Press.

Congressional Research Service. Library of Congress. 1990. "Provisions of the 1990 Farm Bill." 90-553 ENR. Coordinated by Geoffrey S. Becker, Environment and Natural Resources Policy Division, November 16.

Fallert, Richard F., Don P. Blayney, and James J. Miller. 1990. "Dairy: Background for

1990 Farm Legislation." Staff Report AGES 9020. Economic Research Service, U.S. Department of Agriculture, March.

Henderson, Harry W. 1955. *Price Programs.* Agriculture Information Bulletin 135. U.S. Department of Agriculture, Washington, D.C., January.

Marsh, John M. 1988. "The Effects of the Dairy Termination Program on Live Cattle and Wholesale Beef Prices." *American Journal of Agricultural Economics*, vol. 4, pp. 919–28.

Rojko, Anthony S. 1957. *The Demand and Price Structure for Dairy Products.* Technical Bulletin 1168. Agricultural Marketing Service, U.S. Department of Agriculture, Washington, D.C., May.

U.S. Department of Agriculture. 1968. *Dairy Price Support and Related Programs: 1949–68.* Agricultural Stabilization and Conservation Service. Washington, D.C.: U.S. Government Printing Office, December.

_____. Agricultural Stabilization and Conservation Service. 1991. "Milk Inventory Management Report." June 14.

_____. Agricultural Stabilization and Conservation Service. 1994. "ASCS Commodity Fact Sheet: 1993–94 Dairy Price Support Program." June.

U.S. Department of Agriculture. Economic Research Service. 1994. *Dairy Yearbook, 1994.* SB-889. Washington, D.C., August.

U.S. General Accounting Office. 1985. "Effects and Administration of the 1984 Milk Diversion Program." Report to the Congress of the United States. GAO/RCED-85-126. July.

_____. 1990. *Federal Dairy Programs: Insights into Their Past Provide Perspectives on Their Future.* Report to the Chairman, Committee on Agriculture, Nutrition, and Forestry, U.S. Senate. GAO/RCED-90-88. February.

Chapter 9

Local and State Milk Regulation

Extensive regulation for milk marketing and pricing exists at the federal level, including legislation for dairy cooperatives, the price support program, and federal milk marketing orders. This is not the limit of regulation on milk and dairy products, however. Milk, due to its perishability and susceptibility to food-borne diseases, is inspected at the local and state level for wholesomeness. From the cow's udder to the consumer's lips, milk and dairy products are inspected at every step to ensure their safety for human consumption. In addition to inspection, various states have their own regulations regarding the marketing and pricing of milk. Some of these state programs work in conjunction with federal milk marketing orders, while others operate in their absence.

Sanitary Regulation and the Pasteurized Milk Ordinance

Milk has the potential to serve as a vehicle for disease and has, in the past, been associated with disease outbreaks of major proportions. In 1938, milk-borne outbreaks constituted 25 percent of all disease outbreaks due to infected foods and contaminated water (U.S. Department of Health and Human Services 1993). Today, milk and fluid products are associated with less than 1 percent of all such reported outbreaks. Yet outbreaks do still occur. As recently as 1986, a single Chicago milk plant caused an epidemic of listeria and salmonella that affected 18 states, caused an estimated 200,000 illnesses, and resulted in 18 deaths (Long and Drucker 1995). The sanitary regulation of milk is serious business.

The concern is that some animal diseases can be transmitted to humans via milk. Conversely, humans can introduce microorganisms and other adulteration to milk during its production, hauling, processing, and delivery. In any event, there is a significant risk to public health when milk is not handled properly. Campbell and Marshall (1975) outline the major disease-producing microorganisms that are associated with milk. Tuberculosis and brucellosis are the most significant diseases of dairy cattle that can be transmitted directly to humans. Common microorganisms that can infect humans via improperly handled milk include bacteria of the genera *Salmonella* and *Streptococcus* as well as *Staphylococcus aureus* and strains of *S. aureus* that cause bovine mastitis. Adulterants to milk that are of a public health concern are antibiotics that are injected in cows and present in milk, pesticides that accumulate in the fat of cows and are excreted in the milk, radio nuclides from radioactive fallout, and aflatoxins and other mycotoxins that are chemical toxins produced by molds that may be present in animal feedstuffs and transmitted to milk.

A drastic reduction in milk-borne outbreaks occurred this century and can be attributed in large part to the standards set by the Grade A Pasteurized Milk Ordinance (PMO) of the U.S. Public Health Service. Public Health Service activities in the area of milk sanitation began at the turn of the century with studies on the role of milk in the spread of disease. Exhaustive epidemiological studies completed in 1908 firmly established the role of milk in the spread of gastrointestinal and other diseases. They showed that milk-borne disease in the United States was an important public health problem and that effective public health control required the application of sanitation measures throughout the production, handling, pasteurization, and distribution of milk.

The Public Health Service began milk sanitation investigations in 1923 and established the Office of Milk Investigation. A study of state and municipal milk regulations revealed that regulations varied significantly. Some regulations had little or no relevance to milk sanitation and public health, others contained impractical or unnecessary requirements, and others differed in their treatment of the same item of sanitation.

In 1924, to assist states and municipalities in initiating and maintaining effective programs for the prevention of milk-borne diseases, the Public Health Service developed a model regulation, known as the Standard Milk Ordinance, for voluntary adoption by state and local milk control agencies. In order to provide a single interpretation of this ordinance, an accompanying code was published in 1927 that provided the administrative and technical details required to meet compliance. This code was drafted for direct adoption into municipal law. This model regulation, now titled the Grade A Pasteurized Milk Ordinance—1993 Recommendations of the U.S. Public Health Service/Food and Drug Administration, represents the fifteenth revision since 1924 and incorporates new knowledge into public health practice. This model regulation, like its predecessors, was developed with mutual assistance from milk sanitation and regulatory agencies at every level of federal, state, and local government including producer, plant operator, sanitarian, and equipment manufacturer associations and educational and research institutions.

The PMO is therefore a national standard for the sanitary control of Grade A milk and dairy products. It provides detailed recommendations at every stage of production, processing, pasteurization, and distribution of milk and milk products. It also acts as a model milk regulation designed to assist states and municipalities in initiating and maintaining effective programs for the prevention of milk-borne diseases. Another objective of the PMO is to facilitate the shipment and acceptance of milk and milk products of sanitary quality in interstate and intrastate commerce.

As urbanization accelerated in the 1940s and as the interstate highway system developed, concern grew about preventing contamination of highly perishable dairy products shipped in interstate commerce. In order to protect the milk supply, many states sent their inspectors to inspect milk producers in other states. For example, prior to 1980 the Missouri State Milk Board regularly sent its milk inspectors to Texas, Arkansas, Illinois, Wisconsin, Kentucky, Iowa, Nebraska, and Kansas. In fact, Missouri milk inspectors traveled all the way to Washington state in the 1960s in order to inspect 50 producers who shipped milk directly to Missouri. Clearly there was a tremendous duplication of inspections.

The National Conference of Interstate Milk Shipments was developed in 1950 to allow for the free interstate movement of milk and to avoid overlapping inspection services. The conference is composed of representatives from state and local regulatory agencies, the dairy industry, and the Food and Drug Administration. It meets every two years and deliberates on problems that affect the processing and distribution of dairy products across state lines. It is during this conference that the PMO is revised. Suggested changes are reviewed and deliberated by state delegates to the conference. Any actions proposed by the delegates must be approved by the Food and Drug Administration.

The conference has resulted in a cooperative agreement negotiated among all 50 states, the District of Columbia, and U.S. Trust Territories that participate in this program. The PMO is required for certification of interstate milk shippers, which allows for the free movement of milk across state lines.

Manchester (1983) reports that at the time of the Great Depression, local sanitary regulations were widely used to erect barriers to the movement and marketing of milk. These included refusal to inspect farms or plants beyond the existing milkshed, outright exclusion by ordinance of milk from plants outside the city limits or a prescribed radius, excessive fees for inspection at a distance from the city, and many others. Such barriers did not have long-lasting effects on milk marketing in the 1930s since milk marketing was largely local in nature. This began to change, however, after World War II when milk began to be marketed beyond the local milkshed. Alternative inspection programs and requirements, and varying standards of identity, impeded the movement of milk between markets. Many municipal regulations were challenged in the courts and statehouses in the 1950s and 1960s. Those that were not justified on sanitary or quality grounds were revised. For example, in a historic case, *Dean Milk Company v. City of Madison*, 1951, the U.S. Supreme Court held unconstitutional a provision requiring pasteurization to occur within five miles of Madison (Baumer 1993). The city of Madison, Wisconsin, clearly erected a trade barrier in order to protect the local dairy industry from competition from outside the city. The PMO was designed and modified over the years to break down these barriers in the movement of milk by creating a single ordinance that could be adopted uniformly by all municipalities and states.

Thus sanitary regulation for fluid milk production and marketing essentially moved from the municipal to the state level in order to ensure uniform measures across municipalities within states, and to ensure reciprocity across states to facilitate the interstate movement of milk. In order to allow for intrastate milk movement most states have adopted laws that forbid a municipality from having regulations that are more stringent than state requirements. A provision within the PMO encourages states and municipalities to accept milk from other states and jurisdictions that adopt the PMO and thus have standards that facilitate interstate milk movement.

The promulgation and enforcement of sanitary regulations are carried out in most states either by a state milk board or by the state department of agriculture and/or health. These state regulations require that milk meet PMO standards. For example, in Missouri, the state milk board, which was formed in 1973, has the responsibility to enforce state regulations pertaining to sanitary standards for milk and milk products. Prior to the formation of the state milk board, milk inspection was carried out primarily at the municipal level, with 30 separate milk inspection systems, and by the Missouri Department of Agriculture. The state milk board consists of 12 members who are nominated by the director of the Department of Agriculture and appointed by the governor with the advice and consent of the Senate. Four members are representatives and active members of the staff of four local health jurisdictions. Four are Grade A milk producers. One member represents dairy processors. One member is a consumer at large. The remaining two members are the directors of the Departments of Health and Agriculture, or their designated representatives.

The Missouri State Milk Board carries out its inspection program via contractual agreements with political subdivisions of the state. Those political subdivisions are the public health offices of St. Louis County, Springfield, and Kansas City. Each cooperator provides lab services and hires sanitarians to carry out the necessary inspections. Thus the creation of the board has reduced the number of inspection services from 30 in 1973 to just 3 today, thereby reducing the cost of milk inspection. The state milk board has adopted the model regulations of the PMO and enforces such regulations on all milk produced and marketed in the state. The program is funded by a combination of inspection fees and general revenue.

There is a tremendous economic incentive for dairy farmers to adopt the more stringent regulations required by the PMO. Farmers who do so are issued permits that allow them to ship their milk as Grade A milk and receive much higher prices for it. By setting higher milk prices for "fluid-eligible milk," federal milk marketing orders (Chapter 6) create a pricing incentive for producers to meet the higher sanitation requirements of Grade A milk. Only Grade A milk is fluid-eligible and can be used for bottled milk. Excess Grade A fluid-eligible milk can also be used for manufacturing purposes, but Grade B milk can only be used for manufacturing purposes. The pricing incentive was created in order to allow dairy producers to pay the higher production costs associated with

Grade A milk production. Campbell and Marshall (1975) note that such requirements include

1. Prohibition of the addition of water to the milk supply.
2. Freedom of milk from antibiotics, pesticides, and excessive quantities of radio nuclides from nuclear fallout.
3. Exclusion of colostrum and of milk from diseased animals.
4. Proper construction and maintenance of the milking barn, milk room, and related facilities.
5. Safe and adequate water supply for cows and for cleaning the milk room.
6. Sanitary waste disposal with emphasis on adequate treatment of human wastes.
7. Milk utensils and equipment properly constructed, cleaned, and sanitized.
8. Milking accomplished with minimum risk of contamination from animals, equipment, and personnel.
9. Milk cooling after harvesting from the cow's udder within a prescribed time to a temperature that will satisfactorily resist microbial growth.
10. Control of insects and rodents.

The PMO has a tremendous effect on the processing and distribution of Grade A dairy products. It stipulates how milk shall be legally pasteurized, handled, and processed and the standards of identity the products must meet in order to be legally sold as Grade A dairy products. The PMO, for example, sets minimum standards for solids and fat in fluid milk. At the processor level, the PMO requires inspection of all legally timed and sealed equipment. Sanitation inspectors also thoroughly review the plant. All conveyances for handling and shipping milk are inspected annually. And milk truck drivers' procedures for sampling and measuring milk must be regularly monitored.

The PMO applies to Grade A milk only. Grade B milk is inspected and regulated under a different system that may vary from state to state. Most states, but not all, adopt the USDA's suggested requirements for Grade B milk inspection and sanitation that are in the Code of Federal Regulation, Title 21. The USDA surveys about 37 states every two to three years and reports percentage of compliance. Most states use this information when considering reciprocity with other states for out-of-state Grade B milk shipments.

State Milk Regulation

State control of milk marketing has its origins in the Great Depression. As discussed in Chapter 6, the classified pricing system broke down as milk production increased, fluid milk consumption decreased, and prices fell due to the collapse of the economy. While federal efforts were underway to support national farm prices, many states intervened to create a safety net for farmers. Federal milk marketing orders, which were created in the 1930s, did not have widespread support until the 1950s. Also, because fluid milk markets at the time were purely local in nature (both the market and the milkshed), there was a tendency to control the marketing of milk at the state level. Manchester (1983) notes that the police powers of the state were given clear-cut authority to regulate milk marketing directly. Thus despite the emergence of federal milk marketing orders, the states could and did issue regulations specifying minimum and maximum prices and other terms of trade.

Wisconsin passed the first state milk control law in 1932. The New York legislature passed a state milk control law in April 1933 after an intensive review of milk marketing by a joint legislative committee that recommended legislation (Kling 1993). By the end of 1933 nine northeastern states were fixing producer prices, and seven retail prices.

Most state milk control legislation was aimed at controlling competition since many thought excessive competition was the cause of the Great Depression. Their argument was that competition between handlers drove farm-gate milk prices down. Any effort to stave off such ruinous competition would enhance milk prices for farmers. Milk control agencies and boards were set up to enforce state laws aimed at reducing the number of milk handlers in the market and setting wholesale and sometimes retail prices. They were given the authority to license handlers, issue regulations, make investigations, inspect and audit handlers and plants, require records, hold hearings, and initiate legal enforcement actions (Kling 1993). While state milk control has declined in many states due to numerous court challenges, it is still enforced in a handful of northeastern states to this day.

Major Forms of State Milk Regulation

While state milk control has taken on unique forms in many states, there have been seven major forms of regulation that states have used over time: resale price regulation, price filing, below-cost pricing prohibitions, price discrimination and unfair trade practice regulation, licensing, dating, and producer pricing regulation (Baumer et al. 1986).

Resale price regulation involves setting milk prices at retail, and in some cases at wholesale. Price-setting authority has often been delegated to state milk boards. The objective of resale price regulation clearly was to limit competition between handlers. But such state legislation also led to higher milk prices and slower rates of innovation in milk processing and innovation. Resale pricing regulations at both the producer and retail levels peaked in the 1930s and declined significantly by the 1980s. According to Krueger (1986), 18 states had resale price regulations and 21 states controlled producer prices in 1937. Those numbers dropped to 6 and 13 states, respectively, by 1986.

Price-filing regulations required processors, distributors, and retailers to file prices with the state milk board or commission. These prices were generally made public. The regulations were violated if a firm sold dairy products below the filed prices. While applicable to both retail and wholesale prices, these regulations were generally applied to wholesale prices.

Prohibitions on pricing milk below cost are still prevalent in many states. Twenty states in 1985 still had below-cost pricing regulations on the books (Baumer et al. 1986). Below-cost regulations were adopted in many states during the 1950s when mom-and-pop grocery stores and small dairy processors faced intense competition from large grocery store chains and dairy processors. The mom-and-pop stores in particular charged that the retail chain stores were using milk as a loss leader (a product priced below cost) in an effort to drive them out of business. Below-cost pricing prohibitions were initiated to help reduce this occurrence.

Price discrimination and unfair trade practice regulation prohibited price discrimination, rebates, gifts, interest-free loans, and furnishing equipment by processors and distributors to retailers. In the past, processors and distributors attempted to sidestep resale price regulations and below-cost pricing by offering retailers special deals to lower the effective price.

Licensing was required by many states for milk processors, handlers, dealers, and distributors. Licensing requirements had been used in some states to effectively limit "ruinous or destructive" competition. Applications were denied if they were deemed not in the public interest. In addition, when public hearings were involved, application by a prospective licensee was often contested by those with existing licenses. They argued that the market was already well served and that any new licenses would set off destructive competition that would be fed back to producers.

Dating regulations required milk to be on a retailer's shelf within a specified time after milking or processing. In more recent years, dating regulations set pull dates after which milk had to be removed from the grocery store shelf. While the current objective of dating regulations is to ensure that consumers receive fresh milk, these regulations had been used in the past as a method to discriminate against distant, often out-of-state, milk since local milk could more easily meet the dating regulations. Modern milk trucks and the interstate highway system have removed this advantage for local milk.

A small number of states to this day continue to regulate producer prices and operate state milk marketing orders. For example, Kling (1993) cites that in 1993 only five states fixed prices at retail levels (Maine, Montana, Nevada, North Dakota, and Pennsylvania) and four fixed prices at wholesale. A number of other states still fix producer prices, sometimes in excess of federal order prices in adjoining regions. New York state has a western New York state market order, which is patterned after the federal orders. California, of course, has its own state order that establishes classified pricing and uses a quota system to distribute the proceeds of Class I milk sales.

A major problem with state orders, particularly those that use base plans that set a higher price for some classes of milk, is how to handle out-of-state milk. In the past, complex schemes that involved allocation provisions and compensatory payments were used to protect in-state producers from out-of-state milk. For those states that do have producer base plans, it is often helpful to have geographical boundaries to protect in-state producers, as California does. According to Donald Kullman, vice president of Prairie Farms Dairy, "California is figuratively an island in many ways ... with an ocean on the west, mountains on the east, Mexico on the south, and unfriendly co-ops to the north."

Legal Challenges to State Milk Control

State milk control legislation has been rich fodder for court cases since its inception. The problem lies in how state laws handle out-of-state milk. If legislation is enacted in a state to effectively raise the price of milk, then processors have an economic incentive to seek out alternative sources of cheaper out-of-state milk. State milk control regulators then will need to seek alternative methods of erecting barriers in order to keep this "foreign milk" out. This often involved some form of tax, which is a violation of the Interstate Commerce Clause (ICC). The commerce clause of the U.S. Constitution specifically delegates to the federal government the responsibility for regulating interstate commerce (Baumer et al. 1986). The ICC is often used to strike down state barriers to interstate trade in milk. Hence many courtroom battles over state milk control laws focused on the issue of the ICC.

There were a number of cases in New York that adversely affected state milk control legislation across the country. Just after New York state enacted its price control legislation, it was challenged in court by a New York grocer by the name of Nebbia in Rochester, New York. The case went to the Supreme Court in 1934, where it was reaffirmed that state price fixing was indeed legal. This right was reaffirmed in a 1937 Virginia case, *Highland Farms Dairy v. Agnew,* whereby the Supreme Court reaffirmed the power of the state to fix and enforce minimum resale prices within the state without regard to source. In 1935 another New York case was decided by the Supreme Court. In that case, *Baldwin v. Seelig,* New York state milk regulators required Seelig to pay the New York Class I price for milk brought in from Vermont, including the cost of transportation. This regulation effectively discriminated against out-of-state milk and placed an undue burden on interstate commerce. This case hastened the demise of New York's resale price-fixing effort in 1937 (Kling 1993). The 1949 case *Hood and Sons v. Du Mund* involved the New York state law that required distributors to be licensed. In this case, the applicant/distributor Mr. Hood had the burden of showing that the area in New York he wished to serve was not already adequately served by existing distributors and that a new entrant into the market would not result in destructive competition. As a result of this case, the New York licensing law was struck down as being unconstitutional (Baumer 1993).

In 1951 the case *Dean Milk Company v. City of Madison* centered on this issue of discriminatory enforcement of health laws. Milk authorities in Madison, Wisconsin, refused to allow milk produced more than 25 miles from the city to have a Grade A rating. This was successfully challenged by Dean Milk Company, which shipped milk into Madison from Illinois. The court invalidated the ordinance on the ground that it plainly discriminated against interstate commerce. In another case of discrimination, *Polar Cream and Creamery Company v. Andrews et al.*, the right of Florida state milk control legislation to require all out-of-state milk to receive the lowest class price for milk was successfully challenged. The down allocation provisions of out-of-state milk effectively raised the price of milk for in-state producers and lowered it for out-of-state milk.

In general, the courts have been hostile toward state milk control legislation that attempts to raise milk prices for in-state producers and seeks to keep out or discriminate against milk from out-of-state producers. According to Baumer (1993), the only solution for states that want to raise producer prices is to have control over supply through base plans. Through such plans a higher price is generally paid for milk produced within a producer's base, and a much lower price, often below production costs, for milk produced in excess of the base (overbase milk). Baumer also suggests that it helps if the state is geographically isolated, as are Oregon, Washington, California, and Arizona. But such base plans must allow for new entrants or they may be constitutionally vulnerable to the commerce clause.

State Milk Marketing Orders

At present there are only nine states that have some form of state orders: Virginia, Pennsylvania, New York, New Jersey, Nevada, North Dakota, Maine, Montana, and California. Some of these orders are independent of federal orders, as in California, and others work with federal milk marketing orders, as in Pennsylvania.

According to Bob Yonkers, Pennsylvania has a state marketing order that predates the 1937 legislation that created federal milk marketing orders.[1] An autonomous state agency, currently called the Pennsylvania Milk Marketing Board, regulates six marketing areas in Pennsylvania and enforces minimum prices at the farm level, at wholesale (includes fluid products for retail stores and institutions), and at retail (includes fluid products only). These regional orders work in conjunction with two federal orders (numbers 4 and 36). Since 1988, the Pennsylvania Milk Marketing Board has enforced Class I price differentials that have been higher than federal Class I differentials. This gap has been as high as \$1.35 per hundredweight and has dropped to 80 cents per hundredweight in recent years.

By far the oldest and largest state milk marketing order in terms of pooled milk is the California milk marketing order. California's unique order has its origins in the Great Depression. Diminished markets, low prices, and violence resulted in immediate action to stabilize marketing and prices. California officials applied for licenses under federal milk marketing orders. However, the courts ruled that the federal program could not be enforced in California because milk markets were local and no interstate trade occurred (Butler 1992). As a result, the Young Act was created to establish minimum producer prices in June 1935. This act, while amended several times, is essentially the foundation of California's current milk marketing order.

The California system, like federal milk marketing orders, employs classified pricing and pooling procedures (Boynton 1992). Five classes of milk are used in California. Class 1 milk is for fluid products, yogurt (in-state), sterilized or ultrahigh-temperature (UHT) milk (in-state), and lactose-reduced milk. Class 2 milk is for fluid creams, sour cream, cottage cheese, buttermilk, sterilized creams, yogurt (out-of-state), and UHT milk (out-of-state). Class 3 milk includes all frozen products such as ice cream and frozen yogurt. Class 4a is for milk used for butter and dried milk. Finally, Class 4b is for milk used for cheese other than cottage cheese. These class prices also set minimum component prices for milkfat and solids-not-fat.

The Class 1 price is set by a formula that is weighted by three factors: the cost of milk production, commodity prices as an indication of surplus milk, and real average weekly earnings of manufacturing workers, which reflect the ability of consumers to pay for milk. Minimum prices for Class 4a and 4b are based on formulas that reflect national markets for manufactured products (butter, nonfat dry milk, and cheese). Class 2 and 3 prices are set by differentials that are added to the Class 4a price.

California's marketwide pooling system pays producers on the basis of actual pounds of milkfat and solids-not-fat shipped. Unlike federal orders, however, the California system uses a closed base quota system, which means that the blend price a producer receives depends on his or her quota holdings and total marketings.

In 1969 California adopted statewide pooling of all classified prices. Producers were assigned a production base equal to their marketings during 1966–67 and a quota equal to 110 percent of their Class 1 usage during this period. Overbase milk was any milk marketed in excess of the producers' base. Quota milk receives the highest-class prices whereas overbase milk receives the lowest-class prices. An individual producer's quota holdings have value since milk sold within the quota is worth much more than milk sold outside the quota. Quota holdings are freely tradeable on an open market.

Quotas were adopted in 1969 as an enticement for producers that predominantly shipped to fluid bottlers. These producers did not want to share their Class I sales with other producers. Quotas were issued based wrongly on the assumption that Class I sales would grow over time and that in the future, everyone would share in the higher-valued Class I market. Class I sales did not grow and today California has only 28 percent of its sales for Class I uses; the issue of quotas is still contentious in California.

Pooling occurs each month by estimating the pounds of milk marketed under each use, calculating class prices, and then allocating milk to establish pool prices. The pool price for quota milk is estimated by first allocating Class 1 milk prices and then Class 2 and 3 until the quota is filled. The quota price is determined by weighing class prices by use. The remaining milk is allocated to the overbase milk and uses the lower-valued class prices (i.e., Classes 4a and 4b). Thus, pool prices for quota milk and overbase milk are established. Producers are then paid based on their quota and total marketings. Any marketings in excess of the quota receive the lower-valued overbase pool price.

California made major changes in its pooling procedures in recent years by limiting quota value to a maximum of $1.70 per hundredweight. Therefore, $1.70 per hundredweight times the quota amount is removed from the pool each month and paid to quota holders. The rest is pooled and paid out on nonquota milk.

The California milk marketing order is related to the federal milk marketing system in two respects. First, national markets for manufactured products affect Class 4 prices. Thus, the higher the support price for milk, the higher Class 4 prices are in California. Also, California has the ability to sell surplus butter, nonfat dry milk, and cheese to the Commodity Credit Corporation (CCC) under the federal price support program. Differences between the California and federal system were not considered a problem when California was predominantly a fluid milk state. However, as California began to promote and encourage its manufacturing sector and began to market these products beyond its borders, tensions began to mount.

At the heart of such controversy is the California make allowance. The make allowance is the margin used in calculating Class 4 prices that is used to cover expenses in manufacturing dairy products. The size of the California allowance was larger than the federal make allowance for Class III products. For example, the federal price support program uses a make allowance of $1.37 per hundredweight for cheese compared with $2 in the California system (Jesse and Cropp 1996). California's make allowance is intended to encourage adequate manufacturing capacity in a relatively isolated state. If manufacturing capacity is not available, then milk would have to be hauled out of the state, a very costly proposition.

Section 102 of the 1990 Food, Agriculture, Conservation, and Trade Act of 1990 contained a provision that prevented other states from having a make allowance larger than the federal make allowance. This provision was targeted at the California system (Jesse 1994). Proponents of Section 102 argued that the make allowance allows California manufacturing plants to purchase milk at a cheaper price than other plants in the region and to realize the same prices for finished products. The 1996 Federal Agricultural Improvement and Reform (FAIR) Act appeals Section 102 of the 1990 Farm Bill (Jesse and Cropp 1996). It does, however, partially address the problem. Under the act, no state shall provide a make allowance in excess of $1.65 per hundredweight for milk manufactured into butter and nonfat dry milk or in excess of $1.80 per hundredweight for milk manufactured into cheese.

State Milk Control Legislation in the 1990s

Interest was renewed in state milk control after milk prices plunged to support levels for six straight months in 1990. Many state houses enacted legislation to support prices paid to dairy farmers. The commissioner of agriculture for Massachusetts used emergency powers under an interpretation of existing state law to implement an emergency overorder pricing mechanism in April 1992. This law required all milk dealers to pay a vendor fee on all Class I milk sold in the state. The vendor fee would vary from month to month but was based on one-third of the difference between $15 and the blend price in federal order 21. Approximately one-third of all milk sold in Massachusetts is produced by Massachusetts farmers. The rest is imported into the state. The vendor fee was to be collected by the commissioner and redistributed back to Massachusetts dairy farmers only.

The new administrative order was challenged in court by Westlynn Creamery, which buys most of its milk from outside the state, and two other dairies. The case was upheld at every level of court through the Massachusetts Supreme Court. However, the *Westlynn Creamery and Healy* case, named after the new state commissioner of the Massachusetts Department of Food and Agriculture, was overturned in the U.S. Supreme Court on June 17, 1994, on the basis that it violated the Interstate Commerce Clause.

Westlynn Creamery and Healy had impact well beyond the borders of Massachusetts. In Minnesota, a similar law was passed in 1992 that fixed the Class I price of milk to be no less than $13.20 per hundredweight. Regulations released by the commissioner in 1994 indicated that, in any month that federal order Class I prices fell below $13.20, dealers had to pay the difference both on milk produced in-state and out-of-state. Kling (1993) notes that several dealers motioned for a preliminary injunction because, they said, the commerce clause had been violated. The state commissioner of agriculture, however, contended the premium was a tax protected under the Tax Injunction Act. The case was heard in a federal district court, and an injunction was issued based on the fact that the law impeded interstate commerce. The state issued a second law in 1993 that dealt specifically with the court challenges. This was also challenged in federal district court, and another injunction was issued. The state wanted to appeal it, but since it was similar to the Massachusetts case, the Minnesota attorney general waited for the outcome of that case. When the Massachusetts case was overturned in the U.S. Supreme Court, the Minnesota attorney general chose not to appeal the case.

In Maine, the Dairy Price Stabilization Act was passed in 1991; it was also based on a vendor fee. The Maine Milk Commission established a tax on all packaged milk sold in Maine. The vendor fee became effective whenever the price of Class I milk fell below $16 per hundredweight. The money collected by the tax was then redistributed back to Maine producers only. The program was in effect for three years and returned $7 million to Maine producers. It was challenged by Cumberland Farms in federal district court in Maine. The state prevailed, and the case was appealed to the U.S. Circuit Court of Appeals in Boston. While the case was pending, the Massachusetts law was under review by the U.S. Supreme Court. After the Supreme Court issued its verdict, the U.S. Circuit Court of Appeals issued its decision to knock down the Maine case.

In 1996, regional trade compacts were also being considered in Washington and Oregon and in the Northeast. The Northeast Interstate Dairy Compact had the stated objective of achieving regulatory uniformity among the northeastern states by establishing an interstate commission to regulate dairy farm prices throughout the New England region. To set the farm price, the commission would take formal testimony to determine the "reasonable" costs of production as well as the impact of retail costs on consumers and the regional milk market. All milk consumed in the compact-affected area would be uniformly regulated, even milk produced outside the region and marketed within the region. In essence, the compact would (1) limit the ability of other states to join, (2) allow farmers outside New England who sold milk within the region to benefit from the compact, (3) restrict the interstate commerce commission to regulate Class I milk only, and (4) include protections to ensure that the region does not increase milk production (*Cheese Market News* 1995). The compact could be formed if at least three northeastern states passed legislation and, since interstate commerce would be affected, if sanction was granted by the Congress.

The 1996 Federal Agricultural Improvement and Reform Act contains a provision that allows the secretary of agriculture to approve a Northeast Interstate Dairy Compact if there is compelling public interest in the compact area. The compact would only affect Class I milk sales and would only be in effect until federal orders are consolidated (Jesse and Cropp 1996). The secretary, in fact, did approve the compact and ensuing legal challenges will determine its fate.

NOTES

1. Bob Yonkers, Professor, Pennsylvania State University, personal telephone interview.

REFERENCES

Baumer, David L. 1993. "State Milk Regulations: Back to the Future." *Forces Shaping the Future Dairy Industry*. Proceedings of the 48th Annual Midwest Milk Marketing Conference, edited by Ken Bailey.

Baumer, David L., Richard Fallert, and Lynn G. Sleight. 1986. *State Milk Regulation: Extent, Economic Effects, and Legal Status*. Staff Report AGES 860404. Economic Research Service, U.S. Department of Agriculture, April.

Boynton, Robert. 1992. *Milk Marketing in California: A Description of the Structure of the California Dairy Industry and the Government Programs under Which It Operates*. Dairy Institute of California, September.

Butler, L.J. (Bees). 1992. *Maintaining the Competitive Edge in California's Dairy Industry: Part 1—Organization and Structure*. Edited by Julie Spezia. University of California Agricultural Issues Center.

Campbell, John R. and Robert T. Marshall. 1975. *The Science of Providing Milk for Man*. New York: McGraw-Hill Publishing Co.

Cheese Market News, vol. 15, no. 24. 1995. Highlands Ranch, Colorado: Cahners Publishing, July 28.

Jesse, Ed. 1994. *Section 102: The California Make Allowance Issue*. Marketing and Policy Briefing Paper 46. Department of Agricultural Economics, College of Agricultural and Life Sciences, University of Wisconsin–Madison, April.

Jesse, Ed and Bob Cropp. 1996. *Dairy Title: The Federal Agriculture Improvement and Reform Act of 1996*. Marketing and Policy Briefing Paper 55. University of Wisconsin, April.

Kling, Herbert. 1993. "The Role of State Milk Control." *Forces Shaping the Future Dairy Industry*. Proceedings of the 48th Annual Midwest Milk Marketing Conference, edited by Ken Bailey.

Krueger, Eugene. 1986. *Recent Changes in State Milk Control Practices*. Dairy Situation and Outlook Yearbook. DS-406. Economic Research Service, U.S. Department of Agriculture, July.

Long, Terry S. and Barry J. Drucker. 1995. "Milk Likes You!" *Midwest Hospitality*. The official publication of the Missouri Restaurant Association. Vol. 79, no. 5 (June).

Manchester, Alden. 1983. *The Public Role in the Dairy Economy: Why and How Governments Intervene in the Milk Business*. Westview Special Studies in Agricultural Science and Policy. Boulder: Westview Press.

New York State Legislative Commission on Dairy Industry Development. 1988. *Review of Dairy Regulation: State Milk Control in New York and Contiguous States*. Senator John M. McHugh, chairman. Albany: New York State Senate, March.

U.S. Department of Health and Human Services. Public Health Service. Food and Drug Administration. 1993. *Grade "A" Pasteurized Milk Ordinance*. Public Health Service/Food and Drug Administration Publication 229.

Part III

The Future of Milk Marketing

Chapter 10

International Trade in Dairy Products

The United States had very little experience in the international trade of dairy products prior to the 1980s. U.S. dairy policy was focused mainly on keeping imports out and protecting the integrity of the dairy price support program. The only exports were foreign donations; commercial exports were never seriously considered since U.S. domestic prices were well above international prices.

That began to change, however, after passage of the Food Security Act of 1985. Policy makers revised farm law in order to reduce the cost of farm programs and to improve U.S. competitiveness in international markets. The Dairy Export Incentive Program (DEIP) was created in order to help combat European Community export subsidies and to help U.S. dairy exporters gain practical experience in international markets. The program uses export subsidies (bonuses) to help lower the cost of exports to foreign buyers and thereby makes U.S. products more competitively priced.

The U.S. dairy industry in the 1990s is focusing a great deal of attention on international markets. The North American Free Trade Agreement (NAFTA) has created awareness of potential exports to Mexico, and the successful completion of the Uruguay Round of GATT (General Agreement on Tariffs and Trade) has lowered trade barriers to other countries. As trade-distorting policies are reduced and as the economies of South America and Pacific Rim countries improve, the dairy industry in the United States will likely face improved trade opportunities. To realize those opportunities, however, the United States will have to restructure and refocus more attention on improving production efficiency and meeting the needs of international customers.

This chapter discusses the history of trade policy in the United States as it pertains to the dairy industry. It begins with an overview of the international dairy market. The history of Section 22 import quotas is presented next, followed by a discussion of the mechanics of the Dairy Export Incentive Program, NAFTA, and the Uruguay Round of GATT.

International Market for Dairy Products

Milk is produced in nearly every country of the world. Some countries, like the United States, produce milk from specialty breeds with high milk production. Other countries may use a dual-purpose cow for both milk and meat production.

Major Milk Producers

The top 20 milk-producing countries of the world are presented in Table 10.1. Most of these countries are located in temperate regions of the world, such as North America, Europe, and the former Soviet Union. The largest milk-producing country in the world is the United States, producing 69.7 million metric tons of milk in 1994. Milk production in the United States increased 4 percent over the period 1990–94. In addition to the United States, milk production in the North American countries of Canada (7.7 million metric tons in 1994) and Mexico (11 million metric tons in 1994) are also represented in the top 20 countries in the world.

Europe accounts for a major portion of the world's milk production. Eight of the top 20 milk-producing countries in the world are from the European Union. These include Germany (28 million metric tons), France (25.1), the United Kingdom (14.5), the Netherlands (10.7), Italy (10.2), Spain (6), and Ireland (5.5). Production in most of these countries has declined since 1990 due to reforms in the Common Agricultural Policy. The Eastern European country Poland is also a top 20 dairy country, although its production declined 24.6 percent between 1990 and 1994. The former Soviet Union countries of Russia (42.6 million metric tons in 1994) and Ukraine (18.2) are also top 20 milk-producing countries.

As a general rule, few Asian countries are major milk producers. Yet India is the third largest milk-producing country in the world at 30 mil-

Table 10.1. Top 20 milk-producing countries in the world

Rank	Countries	1990	1991	1992	1993	1994
				(1,000 metric tons)		
1	United States	67,005	66,994	68,440	68,303	69,682
2	Russia	55,715	51,971	47,237	46,800	42,600
3	India	27,500	28,200	29,400	30,600	30,000
4	Germany	31,200	28,916	28,106	28,080	28,050
5	France	26,400	25,700	25,315	24,992	25,120
6	Ukraine	24,360	22,409	19,114	18,376	18,200
7	Brazil	14,500	14,200	15,000	15,300	15,700
8	United Kingdom	14,952	14,503	14,428	14,432	14,486
9	Poland	15,801	14,504	13,060	12,650	11,920
10	Mexico	9,330	10,200	10,700	10,720	11,010
11	Netherlands	11,285	11,047	10,901	10,953	10,750
12	Italy	11,491	11,400	11,300	10,400	10,180
13	New Zealand	7,746	8,122	8,603	8,735	9,768
14	Japan	8,190	8,260	8,581	8,625	8,365
15	Australia	6,435	6,578	6,918	7,530	8,300
16	Argentina	6,400	6,400	7,000	7,400	8,100
17	Canada	7,975	7,790	7,633	7,500	7,700
18	Spain	6,200	6,100	6,000	6,130	6,000
19	Ireland	5,595	5,539	5,588	5,529	5,523
20	China	4,157	4,646	5,031	4,990	5,000

Source: USDA, FAS 1995.

lion metric tons. In addition, Japan produces 8.4 million metric tons and China 5 million metric tons. Another general rule is that countries with a tropical climate are not major milk producers due to heat stress on cows. Yet 3 of the top 20 dairy-producing countries have tropical and semitropical climates: India, Brazil, and Mexico.

Finally, the Oceania countries of New Zealand (9.8 million metric tons in 1994) and Australia (8.3 million metric tons) are also significant milk producers. Unlike the other top milk-producing countries that have large populations, these two countries are relatively sparsely populated and export a major portion of their production.

Major Dairy Exporters

Most dairy products that are traded on the world market are processed dairy products rather than raw milk due to the high cost of transportation. The exception is cross-border trade in cream and raw and condensed milk (e.g., between the United States and Mexico). Hard manufactured dairy products such as butter or butter oil, cheese, nonfat dry milk (NDM), whole milk powder, and casein have up till now represented most of the world's trade in dairy products, as opposed to packaged, value-added products. That is because the former products can be easily stored, are used as inputs for further processing in importing countries, and are often produced and/or exported with government assistance. Trade agreements like the Uruguay Round of GATT will likely result in more trade in value-added dairy products, such as specialty cheese, yogurt, ice cream, and sterilized milk, as barriers to trade are reduced and as the economies and purchasing power of importers grow.

The major dairy product exporting countries are New Zealand, Australia, and Europe (including the European Union, Western Europe, and Poland). The United States has some exports, but it ranks far below these countries. While New Zealand and Australia ranked only thirteenth and fifteenth in 1994 in terms of world milk producers, they are two of the largest exporters of dairy products. One reason for their success is that they are the lowest-cost producers of milk in the world and therefore have some of the lowest dairy export prices. These two countries also have the lowest levels of government assistance of any major exporters (see next section).

With respect to butter exports (Table 10.2), New Zealand is by far the world's largest exporter (268,000 metric tons in 1994). Following New Zealand are Australia (94,000 metric tons), the United States (87,000 metric tons), and the Netherlands (63,000 metric tons). Exports from the Unites States expanded over the period 1990–93 due to assistance from the DEIP, whereas exports from the Netherlands have declined due to CAP reforms (see next section).

The two largest cheese exporting countries in the world are New Zealand and Denmark (Table 10.3). The former exported 138,000 metric tons in 1994 and the latter 135,000 metric tons. Australia and a number of European countries make up the rest of the top 10 countries. While not

Table 10.2. Top 10 butter-exporting countries in the world

Rank	Countries	1990	1991	1992	1993	1994
				(1,000 metric tons)		
1	New Zealand	227	176	222	231	268
2	Australia	51	56	59	78	94
3	United States	31	66	159	161	87
4	Netherlands[a]	83	117	99	98	63
5	Sweden	32	21	17	21	23
6	Finland	37	23	18	18	20
7	France[a]	33	44	53	33	20
8	Germany[a]	95	115	24	22	20
9	Denmark[a]	16	15	14	15	16
10	United Kingdom[a]	4	5	6	6	15

Source: USDA, FAS 1995.
[a]Excludes intra-EU trade.

Table 10.3. Top 10 cheese-exporting countries in the world

Rank	Countries	1990	1991	1992	1993	1994
				(1,000 metric tons)		
1	New Zealand	83	100	107	121	138
2	Denmark[a]	152	155	127	163	135
3	France[a]	76	72	84	96	108
4	Australia	50	63	66	84	97
5	Netherlands[a]	76	82	91	104	95
6	Germany[a]	52	62	60	70	70
7	Switzerland	62	62	67	60	59
8	Italy[a]	27	32	32	40	40
9	Austria	36	30	27	32	36
10	United Kingdom[a]	18	22	18	27	29

Source: USDA, FAS 1995.
[a]Excludes intra-EU trade.

on the list, the United States ranked as the eleventh largest exporter of cheese in 1994.

Australia, New Zealand, and the United States ranked as the top three major exporters of NDM in 1994 (Table 10.4).[1] Prior to 1990 the United States exported very little NDM. Export sales of NDM have expanded greatly since then due to assistance from the DEIP. Other major exporters include the European Union countries of Belgium-Luxembourg, Germany, Ireland, and France, as well as Poland, Canada, and Russia.

The major countries that export whole milk powder (Table 10.5) are New Zealand (306,000 metric tons in 1994), the Netherlands (200,000 metric tons), and France (133,000 metric tons). Other major exporters include the European Union countries of Denmark, Belgium-Luxembourg, the United Kingdom, Germany, and Ireland, as well as Australia and Argentina. The United States exports an insignificant volume of whole milk powder under the DEIP and is therefore not a major player in this market.

Table 10.4. Top 10 nonfat dry milk-exporting countries in the world

Rank	Countries	1990	1991	1992	1993	1994
				(1,000 metric tons)		
1	Australia	97	126	121	131	176
2	New Zealand	178	171	164	114	151
3	United States	10	67	126	147	131
4	Poland	71	85	123	126	85
5	Belgium-Luxembourg[a]	14	39	51	60	58
6	Germany[a]	94	81	102	45	34
7	Ireland[a]	52	35	86	59	33
8	Canada	43	36	30	17	22
9	France[a]	18	12	32	14	21
10	Russia	0	0	6	20	21

Source: USDA, FAS 1995.
[a]Excludes intra-EU trade.

Table 10.5. Top 10 whole milk powder-exporting countries in the world

Rank	Countries	1990[a]	1991[a]	1992	1993	1994
				(1,000 metric tons)		
1	New Zealand			257	262	306
2	Netherlands[b]			190	225	200
3	France[b]			129	107	133
4	Denmark[b]			90	84	92
5	Australia			55	64	74
6	Belgium-Luxembourg[b]			55	70	42
7	United Kingdom[b]			48	33	40
8	Germany[b]			32	35	35
9	Ireland[b]			26	22	28
10	Argentina			1	4	12

Source: USDA, FAS 1995.
[a]Not available.
[b]Excludes intra-EU trade.

Policies of Major Non-U.S. Dairy Exporters

The dairy sector is one of the world's most highly subsidized sectors in many developed nations. These subsidies take two general forms: (1) domestic subsidies that limit production and/or support high internal prices and (2) export subsidies that bridge the gap between the high internal prices and the much lower world prices.

Negotiations on the Uruguay Round of GATT created a need for the ability to measure the degree of a nation's subsidies and compare it with that of other nations. As a result, the Organization for Economic Cooperation and Development (OECD) developed the producer subsidy equivalent (PSE). The OECD defines PSE measures as the value of the monetary transfers to producers from consumers of agricultural products and from taxpayers resulting from a given set of agricultural policies. The OECD defines five categories of agricultural policy measures included in calculations of PSEs:

1. *Market price support*—all measures that affect producer prices.

2. *Direct payments*—measures that transfer money directly to producers without raising consumer prices
3. *Reduction in input costs*
4. *General services*—measures that reduce costs in the long run but are not directly received by producers
5. *Other indirect support*

These five categories can be further reduced to just two categories—those that affect market prices (category 1 in the preceding list) and non-price policy transfers. The former reflects the difference between internal domestic prices and border prices. The latter reflects all nonprice budgetary-financed support.

The OECD (1993, p. 232) expresses the PSE as follows:

Gross total PSE	=	$Q \cdot (P - \text{PWnc}) + \text{DP} - \text{LV} + \text{OS}$
Net total PSE	=	$Q \cdot (P - \text{PWnc}) + \text{DP} - \text{LV} + \text{OS} - \text{FA}$
Unit PSE	=	$\text{PSE}u = \text{PSE}/Q$
Percentage PSE	=	$100 \cdot \text{PSE}/(Q \cdot P + \text{DP} - \text{LV})$

where Q = volume of production, P = domestic producer price, PWnc = world price (reference price) at the border in domestic currency, DP = direct payments, LV = levies on production, OS = all other budgetary-financed support, and FA = feed adjustment (only for livestock products). In its simplest terms, the percentage PSE measures the gap between the high internal price and the lower world price relative to the internal price. It is this gap that is attributed to price-distorting domestic and/or export subsidies.

The OECD reports the net percentage PSEs for member countries (Table 10.6). The country with the highest PSE in 1993 was Japan (89 percent), followed by the Western European countries of Norway (82 percent), Switzerland (81 percent), and Finland (75 percent). The countries with the lowest PSEs, and with domestic prices closest to international prices, were New Zealand (2 percent) and Australia (26 percent). The net percentage dairy PSE for all OECD countries in 1993 was 62 percent, which was the highest of all major commodity groups in the OECD. Clearly dairy is the most highly subsidized and protected of all agricultural commodities in many countries in the world.

Table 10.6. Percentage producer subsidy equivalents for milk in OECD countries

	Average 1979–86	1987	1988	1989	1990	1991	1992	1993
Canada	64	77	69	71	81	82	75	70
United States	61	64	50	50	61	55	53	53
Japan	82	90	86	82	87	87	86	89
Australia	28	29	20	21	34	38	33	26
New Zealand	21	13	4	3	3	2	2	2
EU	51	72	64	57	55	69	66	64
Austria	46	59	52	49	64	65	67	67
Finland	64	74	75	72	74	75	76	75
Norway	77	78	76	75	81	83	83	82
Sweden	65	72	67	66	74	76	69	66
Switzerland	72	83	78	74	84	83	82	81
OECD	58	66	58	56	68	65	64	62

Source: OECD, Directorate for Food, Agriculture, and Fisheries 1994.

Note: Percentage PSE = the net total PSE expressed as a percentage of the value of production at the farm gate.

European Union. The European Union has a complex set of policy regimes authorized by the Common Agricultural Policy (CAP) and designed to ensure a fair standard of living for European Union farmers. CAP was born out of an era when the population remembered starvation during and following World War I and II. Three major elements are support prices and purchases, supply management, and border measures (Blayney and Fallert 1990b). The target price is at the center of CAP. A support system was designed to help ensure that market prices are near the target price. This support system is based on variable import levies, intervention buying, and export restitutions.

Variable import levies are used to limit imports from non–European Union countries. The levy is equal to the difference between the threshold price—or the "at-port" equivalent of the target price—and world prices (BAE 1985). Therefore imports can only enter the European Union at prices equivalent to high domestic prices. This effectively limits imports while insulating the European Union from changes in world market prices. Intervention purchasing by the government essentially sets a floor below which European Union market prices cannot fall. The government stands ready to purchase excess supplies of butter, skim milk, and certain categories of Italian cheeses (BAE 1985). This essentially supports domestic European Union prices for these products near the target price, and implicitly supports the prices of other dairy products.

The support system has resulted in excess supplies of dairy products from its inception. As a result, the European Union has used export subsidies or restitution payments to export excess production onto world markets. It does this by bridging the gap between the high internal prices and the lower world prices. These export subsidies are applicable to a wide range of dairy products (USDA, ERS 1994). The amount of the restitution payment is influenced by the European Union market price and international prices, as well as the cost of moving the product to an export point (the final destination of the product) and the overall export situation facing the European Union (Blayney and Fallert 1990a). Clearly, without these export restitution payments, the European Union would not be able to compete in international markets for dairy products.

The European Union introduced the coresponsibility levy system in 1977 due to mounting surpluses in the late 1970s caused by the support program. Under that system, producers were required to contribute to the costs of disposing of surplus dairy products. The amount of the levy ranged from 0.5 to 3.0 percent of the target price. Later, in 1984, the European Union implemented a milk production quota as part of an overall CAP reform effort. The objective was to limit the amount of stocks moving into intervention and to reduce and control program costs. The amount of the quota was apportioned to member countries.

Canada. The dairy industry in Canada is highly regulated by both federal and provincial policies. These policies include recommending milk prices, supporting the price of milk and dairy products, administering a producer quota system designed to limit production, distributing direct subsidy payments to producers, and undertaking various trade and marketing functions (Griffith et al. 1992). Canadian policy is designed to provide dairy producers with a high internal price. This is accomplished by enforcing a production quota on all Canadian producers, limiting imports of dairy products, and using a support price scheme for butter and skim milk prices to ensure a guaranteed market price. Under this policy regime, supply is restricted to a maximum level determined by the quota. Any production in excess of the quota is penalized by overquota levies. Griffith, Lattimore, and Robertson (1992) note that the consequences of these price support and supply control measures are that Canadian consumers face high domestic prices, taxpayers must pay the cost of direct subsidies, and producers are guaranteed a high return for their restricted supply.

New Zealand. New Zealand is the largest exporter of nonsubsidized dairy products in the world. This is evidenced in Table 10.6; New Zealand has the lowest PSE of all OECD countries. Two unique features of New Zealand have contributed to this feat. First, dairy producers there have the lowest costs of production in the world. This is due to their use of intensive rotational grazing, which reduces housing needs and dependence on high-cost concentrates. This low cost of production translates into low input prices for dairy processors and competitive prices for manufactured dairy products. Second, the New Zealand Dairy Board (NZDB), the sole exporter of manufactured dairy products in New Zealand, has effectively and aggressively marketed New Zealand dairy products in world markets. The board has a fairly large presence in major world markets for dairy products.

In New Zealand, about 90 percent of all milk produced is manufactured into dairy products (Dobson 1990). Of that amount, 85–90 percent is exported, with little or no domestic or export subsidies, by the NZDB into world markets. The NZDB works with dairy farmers and cooperatively owned dairy processors to develop products amenable for exporting overseas. The NZDB purchases the products manufactured by the cooperatives, markets these products overseas, and then returns the net proceeds back to the industry (Blayney and Fallert 1990b). Prior to 1988, farm-gate milk prices were determined by a government organization called the Dairy Product Prices Authority. Prices were set in relation to expectations of, for example, export market conditions and processing costs. Since then, government subsidies and involvement in price setting have been removed, and the NZDB has been given sole authority in setting prices. This was done as part of the restructuring of the New Zealand economy that occurred between 1984 and 1994. The restructuring included a significant deregulation of the agricultural sector (Pasour and Scimgeour 1995).

Due to its small population and dependence on the export market, New Zealand has developed a very sophisticated international marketing team with many overseas offices. As a result of this strong international presence, New Zealand currently represents about 25 percent of world trade in dairy products.

Australia. The Australian dairy industry has undergone a major transformation to a free market environment in the last 25 years. The entry of the United Kingdom (then Australia's largest export market) into the European Union in 1973 forced the Australian dairy industry to become more internationally competitive and to seek new markets. Then, in the mid-1980s, the Australian government granted New Zealand access to the Australian dairy market within the context of a free trade agreement. Because of this newfound competition with the world's most efficient dairy farmers, the Australian dairy industry was restructured with less government intervention in pricing and production.

Renewed interest in world markets became apparent with the introduction of the Crean Plan in July 1986, which linked domestic prices with international market returns. Under the plan, the Australian dairy industry operated a self-financed market support scheme that was designed to raise domestic market prices by a small margin above international prices. This scheme was funded by a levy on all milk produced. These funds were then used to provide an export subsidy (called a "market support payment") on all Australian dairy exports, including the dairy components of processed foods.

Starting in July 1995, Australia introduced a new set of marketing arrangements for manufactured dairy products due to its commitments under GATT (*Dairy Good* 1995). The objective of the new plan is to maintain the level of support to farmers but to provide this support independent of export sales. Under the new plan, farmers pay a levy on milk consumed domestically as drinking milk, and manufacturers pay a levy on all milk used to produce finished products. The money raised is placed in a Domestic Market Support fund and is used to make a Domestic Support Payment to farmers on all milk supplied for manufacturing. The maximum level of industry support under the plan will be phased down over the years and will terminate by June 30, 2000.

Major Dairy Importers

The world's importers of dairy products are a curious mix of developed countries that are major dairy producers themselves and developing countries with no dairy industry. The United States and many European countries have very established dairy industries yet are major importers as well. Also, countries like Algeria and Egypt, which have no domestic dairy industry, import dairy products for domestic feeding programs.

With regard to butter imports (Table 10.7), Russia is clearly the world's largest importer (151,000 metric tons in 1994). While its imports have dropped in half since 1990, it still imports more than double its nearest rival. Russia is also a major butter consumer. Prior to price deregulation in 1992, per capita consumption was around 11 pounds, which compares with just 4.5 pounds in the United States. The United Kingdom is the second largest importer, with 59,000 metric tons in 1994. Surprisingly, the United Kingdom is also a major butter producer and exporter. Almost all of its imports are from New Zealand, a Commonwealth partner. For example, New Zealand exported 58,550 metric tons of butter in 1993–94 to the United Kingdom, which repre-

Table 10.7. Top 10 butter-importing countries in the world

Rank	Countries	1990	1991	1992	1993	1994
				(1,000 metric tons)		
1	Russia	309	237	215	162	151
2	United Kingdom[a]	62	58	52	55	59
3	Egypt	80	35	44	34	30
4	Mexico	1	7	12	18	20
5	Brazil	8	9	3	5	10
6	Ukraine	0	0	7	7	10
7	India	0	0	0	0	8
8	Romania	15	7	12	2	8
9	Switzerland	4	3	5	6	6
10	Netherlands[a]	7	2	1	5	5

Source: USDA, FAS 1995.
[a]Excludes intra-EU trade.

sented 23 percent of its export market. New Zealand was granted a lower levy for an agreed upon amount of butter to be imported into the European Union as part of the United Kingdom's accession into the European Union (BAE 1985; USDA, ERS 1994). Other major butter importers in 1994 were Egypt (30,000 metric tons), Mexico (20,000 metric tons), and Brazil and Ukraine (10,000 metric tons each).

The United States and Japan are the world's largest cheese importers, with 150,000 metric tons and 142,000 metric tons, respectively, imported in 1994 (Table 10.8). U.S. imports are limited due to Section 22 licenses that constrain imports. These imports have generally increased with milk production. Russia, the third largest cheese importer, has seen a major expansion in imports, from 8,000 metric tons in 1992 to 76,000 metric tons in 1994. Other major importers are Mexico, Australia, Canada, and some European countries.

NDM is used by many countries for domestic feeding programs for the poor. It is often purchased in conjunction with butter oil for reconstitution into a nutritious and low-cost dairy beverage. NDM is also used as an ingredient in many processed foods. The largest importers of NDM in the world (Table 10.9) are Mexico (200,000 metric tons in 1994), Algeria

Table 10.8. Top 10 cheese-importing countries in the world

Rank	Countries	1990	1991	1992	1993	1994
				(1,000 metric tons)		
1	United States	135	135	129	145	150
2	Japan	106	122	128	137	142
3	Russia	15	13	8	20	76
4	Mexico	12	15	20	30	35
5	Italy[a]	37	36	35	34	34
6	Switzerland	26	28	28	29	31
7	Australia	21	23	25	25	27
8	Canada	18	21	21	22	21
9	Germany[a]	21	17	18	20	20
10	Sweden	21	23	21	20	20

Source: USDA, FAS 1995.
[a]Excludes intra-EU trade.

Table 10.9. Top 10 nonfat dry milk-importing countries in the world

Rank	Countries	1990	1991	1992	1993	1994
				(1,000 metric tons)		
1	Mexico	288	48	187	200	200
2	Algeria	105	80	102	150	128
3	Japan	81	117	96	73	90
4	Brazil	35	63	14	32	45
5	Russia	14	22	2	20	16
6	Chile	8	8	10	10	10
7	Spain[a]	3	4	3	8	8
8	Peru	10	10	8	7	6
9	Canada	0	1	1	5	5
10	Netherlands[a]	7	5	3	6	5

Source: USDA, FAS 1995.
[a]Excludes intra-EU trade.

(128,000 metric tons), and Japan (90,000 metric tons). Most of Mexico's imports are for social programs and are handled by government agencies. CONASUPO purchases the product in international markets, and these imports are distributed through LICONSA. Other major importers of NDM in 1994 were Brazil (45,000 metric tons), Russia (16,000), and Chile (10,000).

Like NDM, whole milk powder is imported and used both for domestic feeding programs as well as an ingredient in many processed food items. Unlike NDM, which contains milk solids only, whole milk powder also contains milkfat. Therefore it can be conveniently reconstituted at any time with water only. The world's largest importer of whole milk powder (Table 10.10) in 1994 was Algeria (98,000 metric tons), followed by Venezuela (41,000 metric tons), and Brazil and Mexico (both at 35,000 metric tons).

Section 22: Import Quotas and Tariffs

Chapter 8 discusses the dairy price support program. That program was created to support milk prices above a free market–clearing level. In other words, the program was designed to remove surplus dairy prod-

Table 10.10. Top 10 whole milk powder-importing countries in the world

Rank	Countries	1990[a]	1991[a]	1992	1993	1994
				(1,000 metric tons)		
1	Algeria			108	98	98
2	Venezuela			54	54	41
3	Brazil			17	26	35
4	Mexico			25	35	35
5	Peru			19	27	22
6	Chile			23	27	20
7	Egypt			21	17	18
8	Argentina			16	4	5
9	Canada			2	3	3
10	Australia			2	2	1

Source: USDA, FAS 1995.
[a]Not available.

ucts from the domestic market and thus raise the manufacturing grade price of milk to at least the support price level.

A problem with this program was that it would attract imports of dairy products into the U.S. market when international prices were below domestic prices. For that reason, the Agricultural Adjustment Act of 1933 was amended in 1935 to include authority for imposing import quotas. This authority was contained in Section 22 of the Agricultural Adjustment Act of 1935. Import quotas for many years protected the domestic market from overseas imports. They have since been replaced with tariff-rate quotas under the Uruguay Round of GATT.

Section 22 directs the secretary of agriculture to inform the president whenever he or she has reason to believe that imports are rendering a price support or stabilization program ineffective, materially interfering with the operation of these programs, or adversely affecting domestic production. If the president agrees with the secretary's assessment, he or she then directs the U.S. International Trade Commission to conduct an investigation, including a public hearing, and submit a report of its findings and recommendations.[2] The president is authorized, based on such findings, to impose such fees or quotas in addition to the basic duty as he or she shall determine necessary. Additional import fees may not exceed 50 percent ad valorem, and quotas proclaimed may not be less than 50 percent of the quantity imported during a previous representative period, as determined by the president. The president may, however, take action without awaiting the report and recommendation of the U.S. International Trade Commission.

Import quotas were first imposed on dairy products (butter, butter oil, casein, cheese, and dried skim milk) in August 1951 under authority of Section 104 of the Defense Production Act of 1950. At the time, U.S. farm prices were rising sharply, and that attracted imports from Europe. Quotas were used to limit the flow of imports to protect domestic production and the price support program. Quotas on imports of dairy products were used again in 1953 under presidential proclamation (no. 3019 of June 8, 1953). Authority came for the first time from Section 22 of the Agricultural Adjustment Act of 1933, as amended.

By enacting revisions and amendments to Section 22, Congress in effect asserted the right to employ import controls over any GATT obligations. In 1953, the United States petitioned GATT to allow this exceptional treatment for agriculture. Contracting parties to GATT acceded and granted the United States what has since been called a "Section 22 waiver." Blayney and Fallert (1990b) note that this exception resulted in other GATT member countries erecting similar barriers after using the U.S. GATT waiver as a defense.

Section 22 was later modified in 1979 after the conclusion of the Tokyo Round of the Multilateral Trade Negotiations under GATT. Four major changes resulted: (1) U.S. rights to countervail against subsidized quota imports were waived provided the import prices do not undercut U.S. prices, (2) European Community access for cheese imports was restored to amounts somewhat equivalent to what the European Community supplied prior to the 1974 countervailing agreements, (3) new quota levels were established for the Oceanic countries and Finland, and (4) increased controls were placed on imports by bringing under absolute quota most of the cheese that previously entered quota-free (USDA, FAS 1988). Until the Uruguay Round of GATT, Section 22 quotas have been essentially fixed since the Tokyo Round negotiations at approximately 2.24 billion pounds (milk-equivalent, milkfat basis).

The quota system operated on the basis of licenses issued by the secretary of agriculture. The purpose of the licensing system was to allocate the quotas for various dairy products in a fair and equitable manner among importers and users. A major portion of the quota was allocated to historical licensees that were actually in the business of importing during a representative period of time that the quota was based on. A smaller portion was allocated to nonhistorical or new licensees. As a result of the Trade Agreements Act of 1979 and Presidential Proclamation 4708, cheese import quotas increased by 53,000 metric tons to 111,000 metric tons.

In addition to Section 22 import quotas, tariffs had also been applied to dairy product imports. Blayney and Fallert (1990b) quote an ad valorem tariff of 6.25 percent for cheese and a straight tariff of 5.6 cents per pound on butter. These tariffs, however, paled in comparison with the impact of the quota system in terms of effectiveness of limiting imports.

Dairy Export Incentive Program

The Dairy Export Incentive Program (DEIP) has its origins in the Food Security Act of 1985. The program was designed to provide cash or commodity export subsidies to help the United States meet unfair and subsidized competition in the world market from the European Union. A second objective was to provide U.S. exporters with much needed experience in exporting onto the world market. Prior to the DEIP, the United States had very little experience in exporting dairy products outside of foreign donation programs.

The program begins each year by targeting countries for possible DEIP awards. A list of export quantities for milk powder, butter, and cheese and countries eligible for DEIP subsidies are assembled by an interagency committee in the USDA. The committee considers targeted countries and commodities based on four criteria (Bailey 1993):

1. Countering export subsidies and unfair trade.
2. Expanding U.S. exports of dairy products.
3. Minimizing the effect of the DEIP on those countries that do not use export subsidies.
4. Maintaining bonuses at a minimum to achieve export expansion and trade policy objectives.

Once the initial allocation is made, it is up to private traders to make actual sales and to move the products. These traders are eligible to negotiate a sales contract with a targeted country contingent on getting a cash bonus award for each unit sold. This effectively lowers the export price to the buyer to world market–clearing levels. After the sales contract has been negotiated, the exporter can then submit a bid to the USDA requesting the bonus award that would allow the sale to take place.

The Foreign Agricultural Service of the USDA reviews each bid from exporters and then accepts or rejects it based on world prices and the size of the bid. It will not accept a bid that effectively undercuts world prices in a particular market. Also, bids that are too high are rejected since this would raise program costs. Bonus levels are expected to cover the difference between the sales price and a "reasonable" world price in a market and to cover costs of transportation. Once accepted, bids are then awarded by the USDA, and the sale takes place.

The DEIP has had a dramatic effect in increasing U.S. exports (Table 10.11) and tightening up the market for surplus milk. This was especially true after Class IIIa pricing became effective and created a surplus class for milk used to produce NDM. Some of this product was exported via the DEIP. In fact, it was more cost-effective to export NDM via the DEIP than to sell it to the CCC since the cost of the DEIP bonus was less than the cost of purchasing the product.

North American Free Trade Agreement

The United States, Canada, and Mexico negotiated the North American Free Trade Agreement (NAFTA), which created a single market for 360 million people in the North American continent. The trade agreement was signed by the U.S. president on December 17, 1992, and implementing legislation cleared the U.S. House of Representatives and the Senate almost a year later. This historic agreement has been of particular interest to the U.S. dairy industry, which has been gearing up ever since for increased exports to Mexico. That increased trade has consisted of both homogenous dairy commodities such as nonfat dry milk exported under the DEIP, as well as branded dairy products such as fluid milk, cheese, and yogurt.

NAFTA consists of separate bilateral agreements on cross-border trade in agricultural products between the United States and Mexico,

Table 10.11. U.S. dairy product exports under the Dairy Export Incentive Program

	Nonfat dry milk	Butter/butter oil	Cheese
		(1,000 metric tons)	
Actual shipments[a]			
1992	113.1	23.4	3.2
1993	117.1	20.4	3.1
1994	118.6	37.9	3.4
Subsidized exports under GATT[b]			
1995–96	108.2	43.0	3.8
1996–97	100.2	38.6	3.7
1997–98	92.2	34.2	3.5
1998–99	84.2	29.8	3.3
1999–2000	76.2	25.5	3.2
2000–2001	68.2	21.1	3.0

Source: USDA, FAS 1994.
[a]Calendar year.
[b]July/June marketing year.

and Canada and Mexico. Most of the rules of the U.S.-Canada Free Trade Agreement on tariff and nontariff barriers continue to apply under NAFTA. The bulk of NAFTA applies to nontariff barriers, tariffs, safeguards for producers, rules of origin, and sanitary and phytosanitary regulations.

The United States and Mexico agreed to eliminate all nontariff trade barriers immediately once NAFTA began and to convert these to quantifiable tariff-rate quotas (TRQ). The TRQ's will allow a certain amount of product (essentially a quota) to enter duty-free. Then, anything over that amount will be subject to an overquota tariff that will be phased out over a 10- to 15-year transition period. In addition, the quota level under the TRQ will increase over the transition period and will be eliminated at the end of it.

Mexican import licenses were a significant barrier to free trade between the United States and Mexico prior to NAFTA. Things began to change when Mexico implemented tariff reductions as a precursor to joining GATT in 1986. These tariffs were immediately eliminated, however, as a part of NAFTA. The United States was guaranteed annual duty-free access into Mexico for 40,000 metric tons of nonfat dry milk and whole milk powder. This 40,000–metric ton quantity will grow by 3 percent per year (compounded) over the 15-year transition period. The initial year's level of imports is about 5 percent less than the average of U.S. exports of milk powder to Mexico during 1989–91 (Dobson 1994). Overquota trade will be subject to a higher tariff based on the "tariffication" of Mexico's import license. This tariff will initially be an ad valorem tariff equal to no less than 139 percent, but it will be reduced by 24 percent during the first six years of the transition period, and then phased out by the fifteenth year. For other products, such as evaporated milk and cheeses, tariffs will be phased out over 10 years.

For Mexico, exports to the United States are still subject to Section 22 import quotas. Mexico will, however, be granted improved access to the U.S. market for its dairy products via a TRQ. This TRQ will allow a small duty-free quantity with tariffs on overquota quantities. For milk powder, the TRQ will initially be for 422 metric tons and will grow at a compounded annual rate of 3 percent. The TRQ will have an overquota tariff based on tariffication of the U.S. Section 22 import quotas. This overquota tariff will be an ad valorem tariff of no less than 78–83 percent. For cheese, the United States will establish an initial 5,550 metric tons under the Section 22 import quota, with an overquota tariff of 69.5 percent. For other dairy products, the United States will establish a number of basket quotas for Mexican dairy products that are currently subject to Section 22 import quotas. The aggregate of these basket quotas for other dairy products (excluding cheese) will be equal to about 5 percent of current U.S. quotas. The overquota quantities will initially be assessed a tariff equal to the 1989–91 average value of the import protection of current quotas.

The quota portion of the TRQs will expand by 3 percent per year, and the tariffs on overquota imports will be phased out over the 10-year transition period. NAFTA does not affect U.S. export programs such as the DEIP. Also, strong rules of origin will ensure that dairy imports from non-NAFTA countries will not be funneled through Mexico.

NAFTA has already had a significant effect on U.S.-Mexican trade in dairy products. A number of export trading companies, brokers, and dairy product wholesalers have set up trade offices in Mexico. These companies have begun to move U.S. dairy products into Mexico as that country is extremely deficient in dairy products. In addition to dairy products, many Mexican dairies have begun to import more dairy heifers from the United States. There are a limited number of large-scale dairy operations located in Mexico that regularly purchase breeding stock.

The decline in the value of the peso relative to the U.S. dollar in 1995 put a temporary halt on expansion of trade activities between the United States and Mexico. Many buyers in Mexico had temporarily stopped imports of dairy products from the United States. However, as the peso began to rally, trade has begun to pick up. U.S. sales of packaged fluid milk, for example, were less price-sensitive than many expected. Many companies moving to Mexico will clearly have to look upon such action as a long-term investment.

Both Canada and the United States excluded their trade in dairy products under the U.S.-Canada Free Trade Agreement (USITC 1993). This provision is continued under NAFTA. This allows Canada to maintain its quota system and high internal milk prices. Without this provision, U.S. exports of fresh milk and dairy products into Canada would be immediate and devastating to Canada; the United States clearly has much lower milk and dairy product prices. This provision under NAFTA has been a bone of contention within the U.S. dairy industry. The National Milk Producers Federation charged that Canada's trade barriers against U.S. dairy products are unacceptable to U.S. dairy farmers, violate three separate trade agreements, and ignore an international trade panel judgment against the barriers (NMPF 1995). Canada refused to eliminate import quotas on yogurt and ice cream after a GATT dispute settlement panel ruled in September 1989 that the quotas were not justified under GATT. This issue with Canada will likely be readdressed again in the near future.

Uruguay Round of GATT

The United States reached a historic trade agreement on December 15, 1993, in signing the Uruguay Round (UR) of Multilateral Trade Negotiations under the auspices of GATT in Geneva, Switzerland. There, delegates representing 117 participating countries formally adopted a 450-page document titled "Final Act Embodying the Results of the

Uruguay Round of Multilateral Trade Negotiations" (NMPF 1994). It was historic in that this was the first GATT agreement that fully included agriculture. GATT-implementing legislation has since been approved by both the House and Senate, and President Clinton signed the trade pact on December 8, 1994. The entire agreement became effective July 1, 1995. It took seven years to negotiate and involved many heated discussions with the European Union, Japan, New Zealand, and many other countries regarding export subsidies, internal support levels, and market access. The major provisions of the Uruguay Round agreement are as follows (USDA, OE 1994):

- *Export subsidies.* Subsidized exports are reduced by 21 percent in volume and 36 percent in budget outlays over a six-year period relative to the 1986–90 base. Countries may phase in export subsidy reductions over the six–year period for any commodity in equal annual increments from 1991–92 levels. Commodities that did not use export subsidies during the base period may not use them in the future.
- *Market access provisions.* All participating countries under the UR must allow minimum access opportunities for agricultural product imports. That level of imports must increase from at least 3 percent of consumption in 1995 to 5 percent by 2000. In order to achieve that goal, all nontariff import barriers are to be converted to tariffs under a process referred to as "tariffication." This process is to make all nontariff barriers more transparent. These new tariffs as well as any preexisting tariffs will be reduced by a minimum 15 percent and, on average, 36 percent over the six-year implementation period.
- *Internal supports.* Total internal support is to be reduced by 20 percent over a six-year period from a 1986–88 base period. Support measures agreed upon as not trade distorting (e.g., conservation measures, crop insurance and disaster assistance, and extension programs) are exempt from reduction.
- *Sanitary and phytosanitary measures.* The sanitary and phytosanitary agreement for the first time enables countries to have GATT rules to bar the use of health-related regulations that are unjustified on scientific grounds and act to restrict trade. On the other hand, these rules ensure a country's right to protect food safety and animal and plant health.
- *Special and differential treatment for developing countries.* Developing countries are allowed smaller reduction commitments of just two-

thirds of the corresponding commitment for developed countries to be implemented over a 10-year period. Least-developed countries are exempt from reduction commitments.

The UR will result in very specific commitments on the part of the U.S. dairy industry, particularly regarding market access and export subsidies. The United States will not have to alter domestic programs, however, since budgetary reductions that have taken place since the base years 1986–90 have already brought U.S. programs into compliance with the UR.

Dairy Market Access

The United States will convert Section 22 import quotas to tariff-rate quotas under the UR. The TRQ will be similar to that used in NAFTA in that it will consist of a two-tier system that establishes one tariff for imports within the quota and another much higher tariff for overquota imports (Dobson and Cropp 1995). Tariffs on the overquota amount will initially provide about the same level of protection that Section 22 import quotas do. Base tariffs on within-quota levels established for 1995 will be reduced 15 percent in equal annual installments over a six-year period as follows:

- Nonfat dry milk: base tariff of 101.8 cents/kilogram reduced to 86.5 cents/kilogram by 2000.
- Butter: base tariff of 181.3 cents/kilogram reduced to 154.1 cents/kilogram by 2000.
- Cheese: base tariff of 144.3 cents/kilogram reduced to 122.7 cents/kilogram by 2000.

Section 22 import quotas for cheese, which totaled 110,999 metric tons prior to the UR, will be converted to a TRQ and increased by 30,992 metric tons to a total of 141,991 metric tons by 2000. This new access will be allocated by country. Of this quantity, 5,550 metric tons are reserved for Mexico in accordance with NAFTA.

In addition to cheese, the United States established TRQs for other dairy products, accounting for 13,700 metric tons of milkfat and 16,100 metric tons of nonfat solids in 1995 (Table 10.12). This amount will grow to 22,785 metric tons of milkfat and 26,825 metric tons of nonfat solids by the year 2000. Small portions of these TRQs are reserved for Mexico under NAFTA. This increase will be allocated to existing Section 22 categories as indicated in Table 10.12.

Export Subsidies

In addition to increased access, the UR also requires member countries to reduce the quantity and value of export subsidies. For the U.S. dairy industry, that implies reducing the DEIP program. By the year 2000, the United States will be required to reduce DEIP sales by 21 percent and 36 percent below the 1986–90 base for the quantity and bud-

Table 10.12. Tariff-rate quotas for other dairy products under the Uruguay Round of GATT

	Quota quantity in 1995	Quota quantity in 2000
Fresh/frozen cream	5,801,600 liters	6,768,500 liters
Evaporated/condensed milk	3,000 mt	7,000 mt
Dried lowfat milk	1,500 mt	5,500 mt
Dried whole milk	550 mt	3,500 mt
Dried cream	100 mt	100 mt
Dried whey/buttermilk	300 mt	300 mt
Butter	4,000 mt	7,000 mt
Butter oil/substitute	3,500 mt	6,100 mt
Dairy mixtures[a]	2,100 mt	4,300 mt
Chocolate crumb	16,000 mt	26,700 mt
Lowfat chocolate crumb	2,123 mt	2,123 mt
Milk replacer feed	7,400 mt	7,400 mt
Ice cream	3,576,112 liters	5,960,186 liters

Source: USDA, FAS 1994.

Note: mt = metric tons.

[a]One hundred tons is set aside for infant formula containing oligosaccharides.

getary commitments, respectively. Some of the reductions (i.e., nonfat dry milk) will begin from 1991–92 average levels. The schedule for reduced use of the DEIP program is reflected in Table 10.11.

Foreign Compliance with GATT

The UR will also require U.S. compliance with major trading partners. In the case of the European Union, it will have to make large reductions in subsidized exports of butter, nonfat dry milk, cheese, and other dairy products. While the percent reduction in European Union subsidized exports is comparable with that of the United States, the UR will lock in a distinct advantage for the European Union since its level of subsidized exports during the base period was much greater than the U.S. level. In addition to reducing European Union subsidized exports, the UR will also require the European Union to establish new TRQs for 104,000 metric tons of cheese and 69,000 metric tons of nonfat dry milk. This will create more opportunities for U.S. exports to the European Union. Also, since the European Union is the world's largest exporter of subsidized dairy products, commitments to the UR will likely tighten world market prices for dairy products and eliminate swings in international prices. In addition to the European Union, other Western European countries and countries such as Japan and Korea will have to improve their access to foreign dairy products.

Many dairy organizations and politicians sensitive to U.S. dairy interests have complained that the UR is on balance a bad deal for U.S. dairy farmers. Farmers Union Milk Marketing Cooperative president Stewart Huber, for example, has been quoted as saying, "Major flaws in the agreement, including continuing high European export subsidies, reductions in the DEIP and advantages given to cheap foreign competition like New Zealand and Argentina, mean American dairy farmers will not have the export opportunities which GATT proponents have claimed" (Carlson 1994). The problem with GATT is that, unlike NAFTA (which provided a clear advantage to U.S. exports to Mexico), the UR will only break down barriers to trade and will only provide opportunities for increased exports, not guarantee them. Therefore, in order to realize those opportunities, the United States will have to restructure its dairy industry in order to become more competitive in international markets.

10/International Trade in Dairy Products

NOTES

1. Most countries in the world refer to nonfat dry milk as "skim milk powder."

2. The U.S. International Trade Commission was formerly the U.S. Tariff Commission; the name was changed by the Trade Act of 1974 (PL 93-618).

REFERENCES

Bailey, Ken. 1993. "Milk Prices Surge under DEIP." *Hoards Dairymen*, May 10.

Blayney, Don and Dick Fallert. 1990a. *Effects of Liberalized Dairy Imports on the Dairy Support Program*. Report in compliance with Section 4504 of the Omnibus Trade and Competitiveness Act of 1988. U.S. Department of Agriculture, February.

_____. 1990b. *The World Dairy Market—Government Intervention and Multilateral Policy Reform*. Staff Report AGES 9053. Economic Research Service, U.S. Department of Agriculture, August.

Bureau of Agricultural Economics. 1985. *Agricultural Policies in the European Community: Their Origins, Nature and Effects on Production and Trade*. Policy monograph 2. Canberra: Australian Government Publishing Service.

Carlson, Gordon. 1994."Dairy Very Upset with Trade Issues." *Cheese Market News*, vol. 14, no. 10.

Dairy Good. 1995. Volume 6, fourth ed. Australian Dairy Corporation, June.

Dobson, William D. 1990. "The Competitive Strategy of the New Zealand Dairy Board." *Agribusiness*, vol. 6, no. 6, pp. 541–58.

_____. 1994. *Implications of the North American Free Trade Agreement for the Upper Midwestern Dairy Industry*. Babcock Institute Discussion Paper 94-1. Babcock Institute for International Dairy Research and Development.

Dobson, William D. and Robert Cropp. 1995. *Economic Impacts of the GATT Agreement on the U.S. Dairy Industry*. Marketing and Policy Briefing Paper 50. University of Wisconsin Extension, January.

Griffith, Garry R., Ralph G. Lattimore, and John C. Robertson. 1992. "Demand Parameters and Policy Analysis for the Dairy Sector." *Market Demand for Dairy Products*, edited by S.R. Johnson, D. Peter Stonehouse, and Zuhair A. Hassan. Ames: Iowa State University Press.

National Milk Producers Federation. 1994. "The Uruguay Round GATT Agreement: Implications for U.S. Dairy Exports." *Dairy Market Report*, vol. 2, no. 2, February.

_____. 1995. *News for Dairy Co-ops*. Vol. 53, no. 2. Arlington, VA, January 16.

Organization for Economic Cooperation and Development. 1993. *Agricultural Policies, Markets and Trade: Monitoring and Outlook, 1993*.

_____. Directorate for Food, Agriculture, and Fisheries. 1994. *Current Situation, Short Term Outlook, and Recent Policy Changes in the Dairy Markets of OECD and Observer Countries*. Group on Meat and Dairy Products of the Working Party on Agricultural Policies and Markets, August.

Pasour, E.C., Jr. and F.G. Scrimgeour. 1995. "New Zealand Economic Reforms: Implications for U.S. Farm Policy." *Choices*, second quarter.

U.S. Department of Agriculture. Economic Research Service. 1994. *International Agriculture and Trade Reports: Europe Situation and Outlook Series*. WRS-94-5. September.

U.S. Department of Agriculture. Foreign Agricultural Service. 1988. *Meat and Dairy Monthly Imports: Handbook on Section 22 Dairy Quotas and Import Licensing System*. Circular Series, Supplement 4-88. April.

_____. 1994. *GATT/Uruguay Round Fact Sheet: Dairy*. June.

_____. 1995. *Dairy: World Markets and Trade*. FD 1-95. April.

U.S. Department of Agriculture. Office of Economics. 1993. *Effects of the North American Free Trade Agreement on U.S. Agricultural Commodities*. March.

_____. 1994. *Effects of the Uruguay Round Agreement on U.S. Agricultural Commodities*. GATT-1. March.

U.S. International Trade Commission. 1993. *Potential Impact on the U.S. Economy and Selected Industries of the North American Free-Trade Agreement*. USITC Publication 2596. Washington, D.C., January.

Chapter 11

Setting Goals for the Twenty-first Century

The objective of this book is to help the reader gain a comprehensive understanding of how market forces, government intervention, and institutions interact to determine milk pricing and marketing in the United States. This chapter summarizes the earlier chapters and assesses how well U.S. dairy programs are functioning. It takes a critical look at how federal milk marketing orders, the dairy price support program, and state orders work together to price and market milk. It then argues for the need to revise dairy policy in order to build for the future and to meet the needs for a global market into the next century.

Chapter 2 presents the theory of milk marketing. That theory is predicated in large part on marketing conditions that existed when federal orders were first initiated back in the early 1930s. Milk was produced in a milkshed that surrounded a city. Only two kinds of milk were produced: Grade A for fluid uses and Grade B for manufacturing purposes. Grade A milk was priced higher than Grade B in order to compensate dairy producers for the higher cost of producing bottle quality milk under strict sanitary conditions. Milk was also produced locally since insulated milk trucks, the interstate highway system, and refrigerated farm bulk tank systems weren't invented yet. In addition, since fluid milk had to be made available at all times of the year, the higher Class I price was also used to encourage sufficient production for year-round use. That of course led to a surplus of Grade A milk that had to be manufactured into storable dairy products.

Marketing conditions today have changed considerably since the early 1930s. Today milk is more often than not produced a long distance from the city center due to sprawling suburban populations that have increased the cost of land. In addition, modern milk tankers regularly move milk up to 1,000 miles. In fact, in 1995, high-quality milk was moved regularly from southeastern New Mexico to Mid-States Dairy, a bottling plant in St. Louis, Missouri, and was trucked as far away as Florida. In addition, more than 96 percent of the nation's milk supply today is Grade A milk. Clearly higher Class I differentials are not justified today to encourage greater production of Grade A milk. The United States is rapidly moving to a single grade of milk. The point is that marketing conditions that created federal orders no longer exist in the 1990s.

Chapter 6 reviews the origins and functioning of federal milk marketing orders (FMMOs). The stated objective of FMMOs was to raise producer income and establish "orderly marketing conditions." FMMOs were clearly enacted by Congress as temporary emergency legislation designed to combat the collapse of the economy and to temporarily eliminate competition among milk handlers. These orders required milk to be priced according to how it was used. Grade A milk was intended for fluid uses, and only surplus Grade A milk was to be used for manufacturing needs. The system of classified pricing back then made sense since milk handlers for the most part received a higher-quality product than the cheese plant did. Also, Grade A milk was worth more than Grade B because it met the new sanitary guidelines for bottled milk laid out by municipalities. Handlers also liked enforced classified pricing because it eliminated price cutting competition that was brutal during the Great Depression. Thus fluid bottlers got something in return for classified pricing and in fact supported it. Pooling requirements also made sense in the local market since Grade A producers had to share in the costs of disposing of surplus Grade A milk for manufacturing uses since they also shared the higher-valued Class I sales. It didn't make sense to pay one producer the lower-valued price of Grade B milk if the cooperative marketed the producer's Grade A milk to the cheese plants. Nor did it make sense to pay the producer's neighbor the higher Grade A price on all of the neighbor's milk marketings. Somewhere someone had to incur balancing costs. Remember, milk could not be transported long distances, and balancing had to be done locally.

Chapter 7 examines the role of dairy cooperatives in the history of the U.S. dairy industry. Clearly dairy cooperatives have played and continue to play a critical role in marketing milk and enhancing producer income. Back in the 1920s when dairy cooperatives were first formed, dairy producers were frequently victimized by unscrupulous milk handlers. These handlers reportedly cheated dairy producers on milk weights and paid them abysmal prices when surplus conditions prevailed. Some dairy producers did not have their milk picked up during the spring flush. Dairy producers were unorganized and had no negotiating strength against large, well-financed proprietary milk handlers. Dairy cooperatives created a critical role in providing a countervailing power in the marketplace. For the first time dairy producers could rely on their cooperatives to not only market all of their milk—a significant problem back in those days—but to do so at a fair price. Cooperatives focused heavily on bargaining with milk handlers on behalf of their members.

Today dairy cooperatives still play a vital role in enhancing producer income by marketing their members' milk and bargaining for overorder premiums. Many cooperatives are also processing bulk dairy commodities and even producing value-added dairy products. Dairy cooperatives today are also very well organized and attempt to favorably influence legislation through prestigious associations such as the National Milk Producers Federation (NMPF).

The federal price support program was reviewed in Chapter 8. Congress experimented with establishing a price support program during the Great Depression. Land O'Lakes Creamery was the first dairy cooperative to become involved in a butter price support program. The dairy price support program was designed primarily (1) to stabilize milk prices and (2) to enhance milk prices above market-clearing levels. Excessive increases in the support price of milk during the 1970s, however, resulted in a mountain of surpluses in the early 1980s that cost taxpayers dearly and expanded the milk supply well beyond market needs. The higher support price and resulting market prices encouraged a lot of dairy producers to get into the dairy business. This of course worsened the surplus problem.

Chapter 9 presents local and state milk regulation. Sanitary regulations for milk were developed by the dairy industry in order to prevent outbreaks of disease since milk was recognized as having the capability to carry food-borne diseases. Sanitary regulations evolved at the local municipal level and are now coordinated at the federal level and enforced at the state level. No one would argue that such regulations are not absolutely necessary to protect the public's health. What is questioned is the growth of state regulations designed to affect the pricing and marketing of milk. Many state legislatures created complicated regulations after the Great Depression to limit competition and to enhance producer prices. Federal orders had not been rapidly disseminated in the 1930s and 1940s, and state orders were created to meet the needs of local milksheds. Many states formed milk commissions that were well staffed and well financed. These commissions controlled how milk was priced from the farm through the retail level. As the transportation system began to evolve, milk began to move into protected markets, and many states attempted to erect barriers through greater regulation. Many court cases were fought over interstate shipments of milk. While today state milk control may be viewed as a throwback to the 1930s and 1940s, it is still used by several states to control how milk is priced and marketed. In fact, the Northeast Interstate Dairy Compact, a component of the 1996 Farm Bill, is largely constructed from anticompetitive ideas that were largely fostered during the Great Depression.

Chapter 10 reviews the potential growth of the world market for dairy products. Domestic and export subsidies from many developed nations are being phased out under the General Agreement on Tariffs and Trade (GATT), and that is providing an opportunity for the U.S. dairy industry. As U.S. exports grow, so also will the opportunity for U.S. dairy farmers to expand production.

What Is Wrong with the Current System

This pricing system that has worked so well over the past 100 years is not adequate to price milk into the twenty-first century. Today dairy farmers no longer receive the benefits of an enhanced pricing regime. This is due to the gradual erosion in the level of dairy price supports. Evidence of this claim is that there were few surplus dairy products moving into government warehouses in the latter 1990s compared with just a few years before. A system that was initially designed to provide enhanced market prices to dairy farmers and stable prices to processors and consumers has been replaced with an unstable system that works at times outside of competitive market forces. The following examples illustrate how the present pricing system falls short of meeting the future needs of the U.S. dairy industry.

Basic Formula Price

The USDA announced in June 1990 that it was conducting an extensive study of alternative pricing mechanisms that would replace the Minnesota-Wisconsin (M-W) method for pricing milk used for manufacturing purposes (USDA, AMS 1991). The problem was that the M-W was based on a dwindling supply of Grade B milk in the Upper Midwest and no longer provided a statistically reliable measure of the market value of Grade A milk used for manufacturing purposes.

Grade B marketings had declined from 33 percent of total U.S. marketings in 1960 to just 8 percent in 1990. After five years of effort the USDA came up with the BFP, or basic formula price. The BFP still relies on a survey of Grade B plants in the Upper Midwest but uses a pricing formula linked to market prices for butter, nonfat dry milk (NDM), and cheese to update the pricing series.

The problem with this new series is (1) it is still based on a survey of Grade B milk, (2) it prices the national milk supply based on unique market circumstances in the Upper Midwest, (3) it provides too much reliance on cheese prices since about 95 percent of all milk used by the survey plants is for cheese processing, and (4) it appears to adjust to market changes more slowly than the old M-W.

The BFP also undervalues the market for Grade A milk used for manufacturing purposes. A report issued by the Agricultural Marketing Service (Nicholson 1996) concludes that "the Class III price [BFP] over the last 11 months has averaged $0.30 per hundredweight less than what might have been expected ... it does appear that the BFP may have fallen short of the level that might normally have been expected, given the cheese prices over the last 11 months."

The Coffee, Sugar, and Cocoa Exchange (CSCE) recently created futures contracts for fluid milk. These contracts were designed to reflect the market value of a tanker load of Grade A milk for delivery to the Upper Midwest. These futures contracts for the first time show true price discovery for Grade A milk for manufacturing purposes. A review of these contract prices clearly shows that the market is willing to pay much more for Grade A milk for manufacturing purposes than the government-surveyed price for Grade B milk when the market is short.

These futures contracts, while thinly traded, show milk price discovery and indicate that the BFP may be undervaluing Grade A milk for manufacturing purposes during certain months of the year.

Class I Pricing

Minimum Class I prices are administratively determined (1) to enhance producer prices, 2) to limit competition between milk handlers in local markets, 3) to maintain local milk supplies, and 4) to provide advanced pricing to bottlers. The Class I prices under FMMOs are currently determined by a formula equal to the BFP lagged two months plus a Class I pricing differential. This system, which worked so well when prices were relatively stable, is sending mixed pricing signals to dairy farmers.

For example, severe heat stress in the summer of 1995 lowered the milk supply just when U.S. schools opened and demand for bottled milk increased. The BFP was slow to respond to a shortage of milk despite a rapid rise in dairy commodity prices. It rose from $11.23 per hundredweight in July to just $12.08 in September. Also, the lagged BFP in the Class I pricing formula for all federal orders actually decreased September 1995 pool prices since it was based on the July BFP, which decreased 19 cents from the month before. Farmers didn't face higher Class I prices until two months later. By then the percent utilization for fluid purposes declined in many orders, and blend prices suffered. As a result, farmers lost millions of dollars in revenue.

In the spring of 1996 the United States experienced a national shortage of milk due to unusually high grain prices and low prices for calves and cull cows. The daily settlement prices for June fluid milk contracts on the CSCE rose from $12 per hundredweight in January 1996 and closed at $15.25 in June. Also, prices for cheese, butter, and NDM rose in response to strong demand and a short supply of milk.

Part of the reason why the milk supply was so short was that the BFP had been slow to respond to changing market conditions. Also, the lag in the Class I pricing formula depressed the value of the April pool for many federal orders. That is because the BFP in February declined 14 cents per hundredweight from the month before. This doesn't make sense. This administrative pricing system is creating greater price volatility for dairy commodities by artificially adding to milk shortages. AA butter prices for example rose from 72 cents per pound in April to

$1.35 in June due to a severe spring shortage of milk. By not sending a clear price signal to dairy farmers through a higher milk price, the system caused farmers to produce less milk than the market wanted. Higher milk prices were needed to pay for higher feed bills. The USDA's milk-feed price ratio, the number of cows on farms, and milk production per cow were all down during the months of January through June. This created fear in the dairy commodity markets and resulted in unusually high and volatile prices for cheese, butter, and NDM. Consumers also were adversely affected by having to pay higher prices for dairy products.

Class IIIa Pricing

A new class of milk pricing called IIIa began in November 1993 to price milk used for NDM production. Many dairy cooperatives produce NDM to balance a local market. These plants use less skilled workers than a cheese plant, which allows them to be closed down quickly. NDM production in some regional markets, particularly in the West, tends to increase whenever there is surplus milk production.

The California order system prices milk used for NDM production (Class 4a milk) based on a formula linked to NDM prices. In the past powder processors in California had a competitive advantage relative to dairy cooperatives outside of the state that paid the higher Class III price for milk used for nonfat dry milk production. Class IIIa pricing in federal orders therefore acted to even the playing field by allowing out-of-state processors to pay the same price for milk as the California processors.

But a relevant question is whether or not this makes any economic sense. Creation of the Class IIIa price has resulted in a significant increase in NDM production. That excess production has ended up in government warehouses or has had to be exported under the Dairy Export Incentive Program (DEIP) at a significant cost to taxpayers. It has resulted in lower prices to dairy producers. And it has insulated powder processors, mostly dairy cooperatives, from competitive market forces and has created an artificial market that made NDM production more profitable. In effect, it has created an economic incentive for cooperative managers to pay their farm members less while increasing the financial position of the cooperatives. And part of the cost of disposing of the resulting excess production has been paid for by the taxpayers via the DEIP program. It might be argued that milk is very expensive to market and cooperatives must bear the burden to balance regional markets. But with Class IIIa pricing there is less economic incentive to market milk as efficiently as possible.

In a competitive market milk would be allocated to its highest-value use. Butter/powder processors would be forced to compete with cheese processors for available supplies of milk. If they refused to pay the same amount, they would lose their supply of milk. Under free market conditions powder production would decrease, powder prices would rise, cheese production would increase, and cheese prices would decline. The drop in cheese prices, however, would be offset if cow numbers were to continue to decline nationally due to the steady consolidation of milk producers.

But reductions in cheese prices, if they occurred at all, would not necessarily adversely affect farm-gate milk prices since the latter would be based on the total value of milk used for manufacturing purposes, not just cheese prices or the value of Grade B milk in the Upper Midwest. In a free market, the value of all dairy commodity prices would affect the value of milk used for manufacturing purposes.

Blend Price of Milk and Federal Orders

The blend price dairy farmers currently receive inadequately reflects changing market conditions. The current U.S. system of classified prices does not adequately transmit proper market price signals to dairy farmers. The market for dairy commodities today reacts to a shortfall in milk production much differently than just a few years ago. That is because there are few surplus stocks of dairy products in government warehouses. The market today is more willing to bid up dairy commodity prices when it senses strong demand and short supply. But those higher prices are often not reflected in a higher BFP. That in turn depresses the blend price. Thus volatility in dairy commodity prices is at times aggravated by the U.S.–administered pricing system that does not send proper signals to farmers to expand milk production.

This ineffectiveness in transmitting proper price signals to dairy farmers is damaging to the industry as a whole. It hurts dairy farmers since it forces more of them from the industry than would otherwise occur. Farmers with high debt loads and production costs in the spring of 1996 were being forced to exit simply because milk prices were not adequate to cover high feed costs. While the market was willing to pay more, the federal order system was not. This type of situation hurts allied industries including the feed industry, veterinarians, and the dairy supply industry. It also hurts processors and consumers since prices for cheese and other dairy commodities can quickly be bid up when milk shortfalls begin to occur. Butter buyers in 1996 were searching for milkfat substitutes to avoid high butter prices.

As the dairy industry continues to consolidate at the farm level (creating fewer and larger farms), these types of shortfalls will continue to develop. An improved pricing system will help slow down these adverse trends by sending dairy farmers proper signals either to expand or contract milk production based on market conditions.

Free Market

Many dairy producers are convinced that they are still receiving artificially high milk prices and that a return to the free market will result in depression-era drops in prices. Some dairy cooperatives are leading the charge that classified pricing and federal orders maintain orderly marketing and enhance producer income. Some in academia perpetuate the myth that any change in dairy policy would result in lower income to dairy farmers. But what would really happen if the United States were to completely deregulate the dairy industry overnight and move to a free and competitive market?

Presented here is a very simple methodology that asserts that a gradual transition to a complete deregulation of the dairy industry over a two- to five-year period would not result in unstable marketing conditions and would not result in lower milk prices. Let's first address whether or not milk prices would fall under deregulation.

The theoretical model of FMMOs illustrated in Figure 2.9 (Chapter 2) suggests that FMMOs result in above–market-clearing prices and surplus production. This production is then removed from the market by the government. But this model does not accurately describe milk marketing conditions in the latter half of the 1990s. Commodity Credit Corporation (CCC) purchases of surplus dairy products under the dairy price support program have been reduced significantly over the past 10 years due to efforts to curb federal spending. As a result, by 1995 only NDM entered the CCC, and that is only because of the distortions introduced by Class IIIa pricing and the higher CCC purchase price for NDM. Those CCC purchase prices were further increased in 1996 with the rise in the support price for milk to $10.35 per hundredweight. Thus one can conclude that farm-gate milk prices in the latter half of the 1990s were in a state of relative market equilibrium. Given this equilibrium, why would anyone expect milk prices under deregulation to fall below these levels? Prices would only fall if there were surplus conditions characterized by government storage of surplus dairy commodities. The hardships dairy producers faced in the past decade to lower these stock levels via a lower milk support price has finally resulted in equilibrium conditions in the market.

The feed situation in the spring of 1996 shows what happens to processors when farmers face inadequate milk prices. Many farmers went broke and processors found themselves short of milk. This was not in the best interests of dairy farmers, processors, and consumers who will have to pay higher prices in the long run. The marketplace can work if allowed to!

1995–96 Farm Bill

Debate on the dairy title to the 1995 Farm Bill has raised the issue of fairness in national dairy policy. Representative Steve Gunderson (R—Wisconsin) became chair of the House Livestock Subcommittee, which was responsible for writing the draft of the dairy title. His constituents in the state of Wisconsin were convinced that they should not be used as a basing point in FMMOs and that the system discriminates against them with low milk prices. As a result Gunderson proposed a dairy title that included creating one FMMO with five regions (California being one region). The first dollar of all Class I sales would be pooled nationally. The idea was to share part of the pricing benefits of federal orders (Class I milk sales and differentials) on a national basis.

Gunderson's plan was soundly rejected by the National Milk Producers Federation in September 1995. As an alternative, NMPF proposed to floor all Class I milk prices at January 1996 levels (a high level for Class I prices) and to increase them an additional amount in all orders that had over a 40 percent Class I utilization. The idea was to appease producers in the Southeast who were facing financial difficulties and to help pay the high cost of importing milk into the region. Gunderson had threatened to propose a complete deregulation of the dairy industry if national pooling was rejected. True to his word he introduced a "freedom to dairy" title that would have eliminated the price support program in January 1996 and federal orders in July 1996. The bill was incorporated into Representative Pat Roberts' (R—Kansas) "freedom to farm" bill, which contained crop titles that would eliminate all planting restrictions and pay farmers transition payments.

The freedom to dairy title would have eliminated the dreaded budget assessments producers were paying and would have provided transition payments equal to 10 cents per hundredweight during fiscal year 1996, and 15 cents per hundredweight in fiscal year 1997, declining to 5 cents per hundredweight in fiscal year 2002. Market administrators would have still been used to gather market information, audit records, and test milk. They would no longer have enforced classified pricing or operated pools. The proposal would also have allowed for contracts to be made between dairy farmers and processors that would have been strictly enforced by the secretary of agriculture. Contracts would in fact have replaced minimum pricing under federal orders. The enforcement provisions would have allowed a market administrator to use federal intervention to enforce a contract if a processor did not live up to the agreement. This would have been quicker and much cheaper than using the court system. The proposal would have continued DEIP up to the maximum extent possible under the GATT agreement, the secretary of agriculture would have provided advice and assistance to the U.S. dairy industry to set up an export trading company, and the National Dairy Promotion and Research Board would have diverted at least 10 percent of its funds to develop and promote the consumption of U.S.-produced dairy products in international markets.

The U.S. dairy industry was obviously not ready to accept deregulation and instead, under the leadership of Representative Gerald Solomon (R—New York), opted for a modification of current law to consolidate FMMOs and to eliminate producer assessments in exchange for a gradual phaseout of the dairy price support program.

Proposal for Federal Order Changes

The 1996 Farm Bill was signed into law by President Clinton on April 4, 1996. The Federal Agricultural Improvement and Reform (FAIR) Act set in motion a procedure to revamp and consolidate FMMOs. The objective was to make them more workable and reflective of market conditions. Funding for FMMOs will end if the industry does not address changes in FMMOs within three years. Thus careful consideration must be given to a gradual move to a less regulated dairy industry within the confines of the FMMO system. The legislation represents a continuation of the gradual processes of moving the U.S. dairy industry toward deregulation.

The FAIR act thus raises the possibility for major changes in the U.S. dairy industry. Richard McKee, director of the dairy division of the USDA solicited industry input for these reforms in a May 2, 1996, letter to interested parties. To his credit, McKee and the USDA wanted to consider all ideas and suggestions for improving the U.S. milk pricing system.

A better proposal than the FAIR act would be to consolidate the number of federal orders in accordance with the act but to eliminate the BFP price formula and to deregulate the market for Grade A milk used for manufacturing purposes. This can be accomplished by fostering the creation of regional auction markets that will bid on at least a weekly basis for Grade A milk used for manufacturing purposes. In those regions with limited numbers of dairy cooperatives and milk handlers, contracts can be used to develop market prices. Class I pricing, overorder premiums on Class I sales, and pooling would be maintained.

A similar concept was initially proposed by James J. Bowe, president of the Coffee, Sugar, and Cocoa Exchange in New York City (Bowe 1996). The CSCE offers futures and options contract trading for fluid milk, cheese, and NDM. Bowe appeared before two subcommittees of the U.S. House Agriculture Committee on May 16, 1996. He proposed replacing the BFP with a weighted average of daily CSCE milk futures contract transaction prices. He further argued that the futures prices are intrinsically related to cash market prices for Grade A milk delivered to the Upper Midwest. In fact, it can be argued that the CSCE fluid futures contracts are more representative of the actual cash value of Grade A milk (during the last two to four weeks of each contract) than is the BFP since the latter is based on a limited supply of Grade B milk in the Upper Midwest.

There are three drawbacks to Bowe's proposal: (1) futures markets will never completely reflect a true cash market, (2) participation in the futures markets is still limited, and (3) the futures markets currently reflect only one market: the Madison district in the Upper Midwest.

A better proposal would be to develop actual regional cash auction markets for Grade A milk used for manufacturing purposes. The fluid futures markets prove that there can be price discovery for a perishable product like fluid milk. Cash auction markets could perform the same task that some fluid spot markets currently do. The only difference is that prices for milk would be discovered in an open format and publicly released. Markets should meet at least once a week and set milk prices for delivery in three to five days. This would function much like the National Cheese Exchange, where a uniform product is traded each week. Adjustments could be agreed upon for component values and milk quality since each tanker load of milk traded would be different. These regional cash markets would coincide with consolidated federal orders and would create multiple basing points for class prices.

Regional Markets

The preceding proposal suggests moving away from one national market for manufacturing grade milk. A national M-W index was constructed in the early 1960s only because the United States had a national price support program. Since only one support price was announced without regional adjustments, one price was needed for manufacturing grade milk. But in reality there are distinct regional markets for fluid and manufacturing grade milk. The Southeast, Northeast, Mid-Atlantic, Corn Belt, Upper Midwest, Southern Plains, Southwest, Northwest, and California are all distinct markets. They vary due to the mix of dairy products that are sold. A minimum of three to five regional auction markets for Grade A milk for manufacturing purposes should be established.

Transportation Costs

One of the eternal problems in the dairy industry is accounting for transportation costs. Under the proposal presented in this chapter, each auction market would need to adequately account for transportation costs when trading milk. Milk would not be delivered to the auction market each day for sale. Rather, it would be delivered from various points within the regional market three to five days after it was traded. A system would need to be set up to account for these costs. One way would be to create a schedule of freight differentials much like that used by the National Cheese Exchange. Prices would be traded f.o.b. to a central location for the region. The buyer would therefore pay all freight costs. The price, however, would be modified by the schedule.

For example, suppose Madison, Wisconsin, becomes a market center for the Upper Midwest. Fluid milk prices would be traded each week f.o.b. Madison. Thus, if a trade was completed between a supplier in western Minnesota and a buyer in northern Illinois, the supplier would face a discounted price for milk, and the buyer would pay the freight. Regional auction markets would therefore create multiple basing points for Grade A milk used for manufacturing purposes. Again, in those regions where there are few buyers and sellers of milk, contracts can be used to replace regional auction markets. These contracts can reach mutually agreed upon prices for milk.

Futures Contracts

These auction markets would be greatly aided in terms of price discovery if used in conjunction with the existing dairy futures contracts. Industry participants concerned over price volatility would simply hedge with fluid futures contracts. Speculators would be attracted into the industry and would supply the needed liquidity. The existence of a true cash market would help industry participants wanting to use the futures contracts since a clear basis would emerge.

One of the problems with using the current BFP and the futures contracts is a predictable basis may not exist since the BFP does not appear to reflect a true cash market price for Grade A milk used for manufacturing purposes. A true cash market price would help to create a more predictable basis relative to the fluid futures contracts. In addition to the fluid basis, the cash markets for butter, NDM, and cheese and the futures contracts for butter (Chicago Mercantile Exchange), NDM, and cheese would ensure that cash prices in the dairy complex are properly aligned. In other words, cash fluid milk prices would be aligned with cash prices for dairy commodities (just as soybean prices are aligned with those for soybean meal and oil). If those prices become misaligned and margins become distorted, someone will make money and someone will lose money.

Classified Pricing and Pooling

Classified pricing and pooling would still function under this proposal as they have in the past. Class IIIa pricing would be eliminated since it represents anticompetitive pricing. A Class III price would be announced each month for each market based on a daily weighted average (on a volume basis) of the local auction market. These regional prices would then be applicable to relevant federal orders.

As in the case of Bowe's proposal, a rolling weighted average price (month to date) would be announced each day as the month develops. For example, prices announced on May 21 would include the settlement price for the day as well as a weighted average price from May 1 to May 21. The final monthly average price would be announced on the last day of the month.

For example, the May Class III price applicable to a new consolidated Upper Midwest order would be announced on May 31 as a weighted average of all milk auctioned in Madison, Wisconsin, and sold during the month. Class I and II prices for applicable orders would be set equal to the *current* Class III price plus new class price differentials based on the new basing point in Madison. Lags in the Class III price would not be used since this leads to distorted prices and sends improper price signals to milk producers. Bottlers and processors would face the same market information that dairy producers did and would also have the same opportunity to contract on the futures markets as other market participants. The new regional auction markets would thus coincide with new regional federal orders and would create new basing points. Class I prices would need to be revamped based on these new basing points. Pooling of classified prices and reporting procedures under federal milk marketing orders would be the same as now.

Conclusions

Dairy cooperatives and federal milk marketing orders were created in the 1920s and 1930s for the sole purpose of aiding dairy farmers. The clear objective was to help organize dairy farmers and to enhance their income. Today one can question whether those gains have been largely eroded. Class IIIa pricing and the BFP result at times in dairy producers being paid below-market values for their milk. That may discourage efficiency in marketing milk. Fluid processors benefit from today's pricing structure since they know well in advance what their raw milk costs are. Dairy producers, on the other hand, are left with all the uncertainty regarding input and output prices.

Dairy producers have already made a very significant transition to a more market-oriented environment. Spending under the Dairy Price Support Program in 1995 will likely be just one-fifth the level of spending under the program when it first originated in 1950 (in actual dollars). Many hardworking dairy farmers have been forced out of business in the process. But that is a process that has been underway in agriculture since World War II and is an unpleasant reality in any capitalistic society.

The twenty-first century holds many opportunities for the U.S. dairy industry if the United States is willing to position itself for it. This would involve undertaking a significant restructuring of the U.S. approach to marketing. Greater resources would need to be allocated to meeting the needs of both domestic and international customers. New research would need to be undertaken to find new dairy products that would appeal to new customers. More marketing economists would have to be hired to find new ways to export and market more U.S. dairy products. More investments should be made in organizations like the U.S. Dairy Export Council, a new organization set up to expand U.S. exports of dairy products.

The rewards of such a bold step would be greater consumption of U.S. dairy products, renewed confidence among dairy farmers regarding their future, and a growing opportunity for younger families to enter dairying.

But a simple and more transparent pricing system should be a major priority in repositioning the U.S. dairy industry. This change would better serve dairy producers, processors, allied industry, and consumers. In the future, a tanker load of Grade A milk will have a clear market value irrespective of government programs. That tanker load of milk will be subject to component pricing and quality adjustments. Steps should be taken to lift the regulatory cloud and to allow the market to set a fair and equitable price to bring forth a sufficient volume of milk to meet the needs of a growing domestic and international market. This would allow the United States to more effectively use the emerging dairy futures markets. And it would allow the United States to become a world-class leader in overseas dairy markets. The present pricing system will not allow the United States to adequately take advantage of the tremendous opportunity to export more dairy products overseas and to expand its production base at home.

REFERENCES

Bowe, James J. 1996. Testimony of James J. Bowe, president, Coffee, Sugar, and Cocoa Exchange, Inc., before the Subcommittee on Livestock, Dairy, and Poultry and the Subcommittee on Risk Management and Specialty Crops of the Committee on Agriculture, U.S. House of Representatives, May 16.

Nicholson, Donald R. 1996. *Marketing Service Bulletin*. Agricultural Marketing Service, U.S. Department of Agriculture, March.

U.S. Department of Agriculture. Agricultural Marketing Service. 1991. *Study of Alternatives to Minnesota-Wisconsin Price*. Washington, D.C., September.

Index

Index

Index

Index

Index

Index

Index